T0136328

Effective Processes for Quality Assurance

Effective Processes for Quality Assurance

Boyd L. Summers

CRC Press is an imprint of the
Taylor & Francis Group, an **informa** business
AN AUERBACH BOOK

CRC Press
Taylor & Francis Group
6000 Broken Sound Parkway NW, Suite 300
Boca Raton, FL 33487-2742

Printed on acid-free paper

International Standard Book Number-13: 978-0-367-17296-1 (Hardback)

Library of Congress Cataloging-in-Publication Data

Names: Summers, Boyd L., author.
Title: Effective processes for quality assurance / Boyd L. Summers.
Description: Boca Raton, FL : CRC Press, 2019.
Identifiers: LCCN 2018061496 | ISBN 9780367172961 (hb : alk. paper)
Subjects: LCSH: Quality control.
Classification: LCC TS156 .S8465 2019 | DDC 658.5/62—dc23
LC record available at https://lccn.loc.gov/2018061496

Visit the Taylor & Francis Web site at
http://www.taylorandfrancis.com

and the CRC Press Web site at
http://www.crcpress.com

Contents

List of Figures		xi
List of Tables		xiii
Preface		xv
Acknowledgments		xvii
Summary		xix
Author		xxi
Chapter 1	Introduction	1
	1.1 Quality Assurance	1
	1.2 Vision	2
	1.3 Mission	3
	1.4 Quality Planning	3
	1.5 Performing Reviews, Audits, and Evaluations	3
	1.6 Record and Report Reviews, Audits, and Evaluation Results	4
	1.7 Quality Management	4
	1.8 Policy	5
	1.9 Quality Engineering	6
	1.10 Driving Innovation to Reduce Costs	7
	1.11 Lean and Agile Practices	8
	1.12 Proactive Approach to Quality Assurance	9
	1.13 Quality Assurance First	10
	1.14 High-Performance Work Teams	11
	1.15 Plan Implementation	11
	1.16 Tasks and Results	12
	1.17 Quality Management Systems	14

1.18 Quality Assurance Teams 15
1.19 Customer Satisfaction 15
Further Reading 16

CHAPTER 2 QUALITY ASSURANCE METHODS 17
2.1 Quality Assurance Methods 17
2.2 Functional Quality Assurance Tasks 18
2.3 Perform Effective Process Evaluations 18
2.4 Consistent Quality Assurance Process and Reviews 19
2.5 Documentation and Required Data Control 20
2.6 Basic Standards 21
2.7 Quality Assurance Stability 22
Further Reading 22

CHAPTER 3 MEASURE BENEFITS OF QUALITY ASSURANCE PROCESS IMPROVEMENT 23
3.1 What Is Needed for Measurement? 24
3.2 Quality Control 25
3.3 Problems with Healthcare Do Need Quality Assurance 26
3.4 Control Charts 27
3.5 Strategic Planning for Quality Assurance 27
3.6 Commitment to Quality Assurance 27
3.7 Software and Quality Assurance Improve Healthcare 28
3.8 Quality Assurance Process Improvement 28
Further Reading 29

CHAPTER 4 QUALITY ASSURANCE PERFORMANCE AND IMPROVEMENT 31
4.1 Progress 31
4.2 Preparation for Quality Assurance Performance 31
4.3 Preparation of Quality Assurance Improvement 33
4.4 Evaluation for Quality Assurance Processes 33
4.5 Quality Assurance Excellence 34
4.6 Building and Maintaining Quality Assurance Teams 35
4.7 Quality Assurance Problem-Solving 35
4.8 Common Quality Assurance Working Framework 36
4.9 Quality Assurance Impacts Business 36
4.10 Ways to Improve Quality Assurance 37
4.11 Summary 38
Further Reading 38

CHAPTER 5 ANSWER QUESTIONS FROM MANAGEMENT AND EMPLOYEES 39
5.1 Techniques Required 39
5.2 Conduct Questions and Answers 39
5.3 Be Aware of Management Mistakes 41
5.4 Strength Should Never Show Weakness 41

5.5 Quality Assurance Processes Support Employees 41
5.6 Relationship between Management and Employees 42
5.7 Quality Assurance Interviews 42
5.8 Summary 43
Further Reading 44

CHAPTER 6 MANAGE RISK MANAGEMENT 45
6.1 Types of Risk Management Projects 45
6.2 Effective Planning for Risk Management 46
6.3 Risk Mitigation and Monitoring 46
6.4 Risk Management Activities 47
6.5 Risk Management for Quality Assurance Process 48
6.6 Summary 48
Further Reading 48

CHAPTER 7 QUALITY ASSURANCE IMPROVEMENT WITH METRICS 49
7.1 Quality Assurance Metrics 49
7.2 Measure Program and Project Performance 50
7.3 Using Metrics for Quality Assurance 52
7.4 Software Metrics 53
7.5 Metrics Used for Success 54
7.6 Corrective Actions with Metrics 56
Further Reading 58

CHAPTER 8 QUALITY ASSURANCE FRAMEWORK 59
8.1 Required Evidence 59
8.2 Requirements and Risk Mitigations 60
8.3 Using the QAF for Companies and Institutions 61
8.4 Benchmarking Process 61
8.5 Benefits of QAFs 62
8.6 Summary 63
Further Reading 63

CHAPTER 9 QUALITY ASSURANCE PROCESS IMPROVEMENT 65
9.1 Process Improvement Direction 65
9.2 Quality Assurance Process Infrastructure 66
9.3 Effective Process Environment 66
9.4 Document Action Plan 67
9.5 Quality Assurance Process Execution 67
9.6 Quality Assurance Representatives 69
Further Reading 69

CHAPTER 10 EFFECTIVE PROCESSES FOR QUALITY ASSURANCE 71
10.1 Success of Programs and Projects 71
10.2 Communication Needs 72
10.3 CMMI Guidelines for Process and Product Improvement 72

10.4 Process Improvement Services 73
10.5 Decision Analysis and Compliance 73
Further Reading 75

Chapter 11 Quantitative Process Performance and Commitments 77
11.1 Goals and Process Capability 77
11.2 Perform Commitments 77
11.3 Verification and Validation 78
11.4 Quality Engineering Knowledge 80
11.5 Quality Assurance Process Performance 80
Further Reading 82

Chapter 12 Understanding the Quality Assurance Direction 83
12.1 High-Level Quality Assurance Direction 83
12.2 Roles and Responsibilities 84
12.3 Effective Methods for Quality Assurance Direction 85
12.4 Balance Processes and Understanding Required Tools 87
Further Reading 88

Chapter 13 Military Aerospace and Defense 89
13.1 Hercules Missile Defense Program 89
13.2 B2-Stealth Bomber Program 90
13.3 F-22 Raptor Program 91
13.4 Advanced Systems Program 92
13.5 Airborne Early Warning & Control Program 92
13.6 P8A Poseidon Navy Program 93
13.7 Quality Assurance Conferences 94
13.8 Quality Assurance Technology 94
Further Reading 94

Chapter 14 Quality Assurance Plans 95
14.1 QA Communication 95
14.2 QA Plans Provide Education and Learning 97
14.3 Improving QA Plans 98
14.4 Summary 100
Further Reading 100

Chapter 15 Quality Assurance for Customers and Suppliers 101
15.1 Customer and Supplier Requirements 101
15.2 Process Control 102
15.3 Internal Audits 103
15.4 Summary 105
Further Reading 105

CHAPTER 16 SUPPORTING SOFTWARE CONFIGURATION MANAGEMENT 107
16.1 QA and Configuration Management 108
16.2 SCM Planning 110
16.2.1 Software Configuration Management 112
16.3 Change Request Management 112
16.4 Change Requests 112
16.5 What is ClearQuest? 114
16.6 Software Builds and Release 116
Further Reading 117
APPENDIX A 119
INDEX 125

List of Figures

Figure 1.1 Quality is important 1

Figure 1.2 Quality Assurance and Documentation 4

Figure 1.3 Flow of quality engineering 7

Figure 1.4 Agile management model 8

Figure 1.5 Process improvements 14

Figure 8.1 Quality Assurance Framework 62

Figure 9.1 Assessment findings 68

Figure 10.1 Process used for business 74

Figure 10.2 Company and business management 75

Figure 12.1 Team members 86

Figure 16.1 Software Configuration Management 108

Figure 16.2 SCM-tool control 110

Figure 16.3 CR defects 115

List of Tables

Table 1.1 Quality Assurance Activities 9

Table 1.2 Peer Review Method 13

Table 2.1 Quality Assurance Processes 20

Table 3.1 Quality Assurance Performance 24

Table 3.2 Adversity and Concerns 24

Table 4.1 Definitions of Quality 38

Table 7.1 Measure Program and Project Performance 51

Table 8.1 QAF Questions 60

Preface

For years I have wanted to write a book related to Quality Assurance. The new book is titled *Effective Processes for Quality Assurance.* Quality Assurance processes, activities, and plans ensure contract implementation is an accomplishment for Quality Assurance to be implemented in companies, institutions, military programs, and successful businesses. I will also discuss the importance of Software Configuration Management and how Quality Assurance is involved and knowing that processes are followed in software development. My previous books, *Software Engineering Reviews and Audits* and *Effective Methods for Software and Systems Integration,* provided the framework and detailed requirements for software Quality Assurance.

Software Engineering Reviews and Audits—Contents

Software Engineering Reviews and Audits will improve individual and company efforts in maintaining a professional setting where quality is developed for profit, cost reduction, control, and service improvement. Performing reviews and audits for Quality Assurance ensures compliance in specified contracts and customer satisfaction for successful companies and institutions.

Effective Methods for Software and Systems Integration—Contents

To develop, operate, and maintain software and systems integration capabilities inside work product facilities and companies, there must be a major discipline for supporting the entire software life cycle (i.e., planning, systems, requirements, design, builds, installations, integration, subcontractors, quality, and delivery to customers) to be complete and understood.

Acknowledgments

I want to thank my lovely and beautiful wife, Jana, for her support while writing all my past software technology books. Jana always provided great input as well. My beautiful daughters, Julie, Jill, Jann, and Jami, also provided support for me to write many of my books and have always been there for me as a father and author. The opportunity to be an employee, contributor, and consultant for companies, institutions, military programs, and successful businesses has made me a team player and to provide engineering and quality expertise. Currently, I am a software and Quality Assurance technology consultant supporting all software and Quality Assurance activities and providing quality expertise and solutions for software, Quality Assurance, and control.

To ask questions for current and future Quality Assurance solutions, my email is bl.summers.consulting.llc@gmail.com. All important information related to this topic will be provided.

Summary

It is critical to understand and implement the disciplines outlined in *Effective Processes for Quality Assurance* prior to deliveries of important quality products inside and outside companies, institutions, military programs, and successful business environments. Chapters in this book are documented to ensure that internal or external Quality Assurance auditors are trained and chartered to partner for Quality Assurance and product requirement compliance through in-house reviews, audits, and evaluations to provide effective oversight.

Author

Boyd L. Summers is a retired Software Engineer for The Boeing Company and is a Software Technology Consultant for BL Summers Consulting LLC living in Florence, Arizona. Summers is an author of two software technology books titled *Software Engineering Reviews and Audits* and *Effective Methods for Software and Systems Integration*. He provides written software articles to software engineering journals and magazines. Topics include system design, software requirements, software design, software testing and evaluation, configuration management, Quality Assurance, process and product evaluations and the applied processes in Agile, Lean, and Six Sigma.

He is a speaker and a board member of numerous software and quality control conferences around the world.

1

INTRODUCTION

1.1 Quality Assurance

In order to develop, operate, and maintain Quality Assurance process capabilities, there must be a major discipline supporting the entire life-cycle (i.e., commitment, consulting, planning, verification, and validation for design, builds, Quality Assurance, and delivery to customers) needs to be completely understood. The critical understanding and the implementation of the right discipline of these methods empower and achieve effective, flexible, and Quality Assurance results. The right disciplines are identified in Figure 1.1.

More than a word, it is a way of life, and Quality Assurance is more important than Quantity "One home run is much better than two doubles".

The Aerospace standards 9100 (AS9100), Standard Aerospace Evaluation (SAE) AS9110, and International Organization for

Figure 1.1 Quality is important.

Standardization 9001 (ISO 9001) are the models for Quality Assurance of all processes, requirements, production, and service. These standards are used when conformance to processes is to be assured by companies during several stages of compliance. The standards given below adopt and deploy the following:

- Companies, institutions, military programs, and successful businesses should comply with AS9100 for Quality Assurance to show system requirement-specific additional requirements for support.
- Providing Quality Assurance for maintenance, repair, or modification services should be required to comply with AS9100 or may select SAE AS9110 to ensure that processes are compliant and working per Quality Assurance disciplines.
- Software companies' products must comply with ISO 9001 for compliance to minimum requirements for the application of AS9100 and ISO 9001 on software programs.
- ISO 9001 requirements for a Quality Assurance management system can be used for internal application certifications and contractual purposes to focus on the effectiveness of meeting and accomplishing customer requirements.

As a software and Quality Assurance consultant, I like to reference ISO 9001 disciplines and requirements during all reviews, audits, and evaluations. Before implementing reviews, audits, and evaluations, it is important to have future Quality Assurance objectives in place.

Management and all team members must provide the planning for Quality Assurance to be carried out to meet the integrity of quality to be maintained when changes are planned and implemented. Programs and organizations must achieve and sustain quality of the product or service produced to meet the customer needs. Confidence of customers will show that quality is achieved in the delivery of all products provided for use.

1.2 Vision

Effective Quality Assurance processes are inclusive for creating all people to work together and establish an inspired future for all companies, institutions, military programs, and successful businesses.

1.3 Mission

Driving the growth of people and all businesses through personal and professional development focuses on disciplined execution and quality. Quality Assurance Audit Steps are as follows:

- Quality Planning
- Perform Reviews and Audits and Evaluations
- Record and Report Reviews and Audits and Evaluation Results

1.4 Quality Planning

At the start of each audit, review, and evaluation period, Quality Assurance auditors prepare for planning by identifying contracts evaluated during that specific and important period for companies, institutions, military programs, and successful businesses.

The identified contracts and processes require the right criteria derived from documentation associated plans, procedures, and work instructions.

1.5 Performing Reviews, Audits, and Evaluations

This step provides the criteria derived from

- contracts
- documented plans
- defined procedures
- work instructions

Performance of reviews, audits, and evaluations ensures activity performance and processes are compliant with approved directions. Performance and processes are important for auditors to make an assessment as to whether implemented and used processes are compliant or noncompliant.

An auditor's performance helps companies, institutions, military programs, and successful businesses to verify, analyze, communicate, and track technical, financial/costs, schedules, contractual, customer, suppliers, and internal/external risks to ensure long-term success.

Interviews with management and employees ensure that Quality Assurance is implemented for compliance, promoting a professional environment. The auditor identifies an issue or opportunity for improvement, as a result of reviews, audit, and evaluations.

1.6 Record and Report Reviews, Audits, and Evaluation Results

Recording and reporting reviews, audits, and evaluation results provides companies, institutions, military programs, and businesses to maintain historical records (electronic or documents) such that they accurately reflect the activities and status (Figure 1.2). Managing configuration and control of all records as required by contract and requirements are retained for compliance and use.

1.7 Quality Management

Quality management ensures that everything is consistent. It has four main components: quality planning, Quality Assurance, quality control, and quality improvement. Quality management focuses not only on product and service quality, but also on the means to achieve it. Therefore, quality management uses Quality Assurance and control of processes as well as products to achieve more consistent quality. Having quality management in place means simply

Figure 1.2 Quality Assurance and Documentation.

having documented paperwork and online instructions, executing with knowledgeable employees, monitoring or measuring, and making continual improvements. The following improvements are to plan and document to deliver results and do implementation by a skilled work force. Always check and act to take actions and continually improve performance.

In order to have quality management implemented, the companies and institutions must be focused, process-based, and improvement-oriented. Say what you do, do what you say, prove it, and improve it. A quality management system can be used for internal application certifications and contractual purposes, and the focus is on the effectiveness of the quality management system in meeting customer requirements and expectations.

Do what you say (Compliance): Follow all procedures and instructions that affect your work. You must say what you do (Documentation): Use current plans, procedures, and work instructions. Prove it (Records): Demonstrate your work in accordance with compliant processes/procedures and provide objective evidence. Improve it: (Business Management/Continual Improvement): Implement change based on the information provided by business management.

If the specification does not reflect the true quality requirements, the product's quality cannot be guaranteed. For instance, the parameters for a pressure vessel should cover not only the material and dimensions but also the operation, environmental safety, reliability, and maintainability requirements.

The quality of products is dependent on the participating constituents, some of which are sustainable and effectively controlled. The processes are managed with Quality Assurance to pertain total quality management.

1.8 Policy

A policy is the key element in a business process, and there are organizational planning and control documentations and/or procedures to support key elements. The significant activities are defined in this book. To conduct a successful business, we should understand the scope of work to be accomplished. A policy provides a mission statement of direction and guidance for companies, institutions, military programs, and successful businesses. Policies are the highest level of authority and are consistent with the visions that should be used to be successful. A very effective policy to review over and over is a policy for Quality Assurance. The policy states that we are the difference, such as

- I am personally responsible and accountable for the quality of my work.
- I acquire/use the necessary tools and skills needed to meet quality requirements.
- I know my objectives and needed process improvement goals.

Organization-level policies provide the necessary means for companies, institutions, military programs, and successful businesses to establish effective processes to be implemented, supporting project development, modification, and/or procurement. Quality Assurance representatives will always ask questions concerning processes to support what quality is allowed to do and at times can be an annoyance because of the many questions asked.

Management and team members must work well together in all activities such as tracking of changes, verification/validation of processes for compliance, and product development.

All development activities are performed in accordance with defined, repeatable, managed, and optimized policies to ensure that all processes are using a company or program standard, which shows that cost parameters are established, documented, and maintained.

1.9 Quality Engineering

In context, quality engineering is associated with analysis, requirements understanding, and the importance of employer and/or consultant capabilities. Interfaces are defined externally and internally to ensure that Quality Assurance is compatible with supporting companies, institutions, military programs, and successful business activities. The Quality engineering process methods are included in tasks or assignments to integrate all disciplines to meet all requirements and expectations. In years of working with military and aerospace programs, technical quality engineering needs are very important.

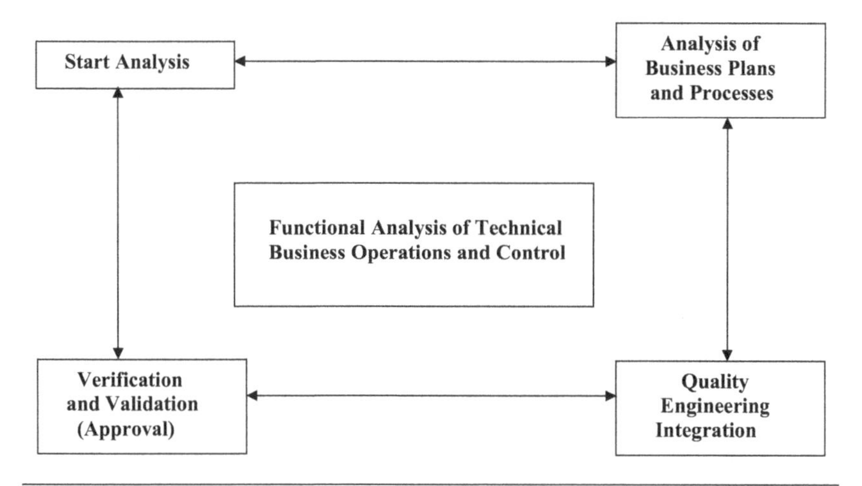

Figure 1.3 Flow of quality engineering.

Quality engineering methods are used for application, setting the ladder for rigorous business techniques to solve complex technical and functional problems. Refer Figure 1.3.

1.10 Driving Innovation to Reduce Costs

Driving innovation will help in reducing costs for companies, institutions, military programs, and successful businesses. Delivery of complex products must have high quality to reduce customer problems and defects. Integration of Quality Assurance processes provides compliant work product management and gap analysis.

The purpose of Quality Assurance is to provide a common operating framework in which best practices, improvements, and cost avoidance activities can be shared, and Quality Assurance responsibilities assigned show results from converging on quality shared by best practices that are improved for process execution and reduction of operational costs.

Quality Assurance personnel must support companies, institutions, military programs, and successful businesses by encouraging a cooperative, proactive approach and ensure compliance through management and team member participation. All results must be reported to what will make sure that all activities and tasks are happening.

1.11 Lean and Agile Practices

Coming from a software and Quality Assurance technology background, I have supported many military and aerospace programs that are Lean and Agile and show a competitive advantage. By implementing these two principles, practices and development deliveries of products to the customers will show Quality Assurance has been applied and have fewer defects.

The definition of Lean is a new concept in the software world. Lean principals establish clear priorities by getting rid of bad multitasking, focus, and not finishing the task assigned to an individual within business companies and military and aerospace programs. Lean principals will eliminate the release of software being late and require an early delivery. One must prepare, start, finish, and use checklists to prevent software defects and risk. Teams will face and resolve issues on a timely basis and drive daily software execution and quality products.

Applying the Agile Management Model per Figure 1.4 implements software development, supports many initiatives, and provides military and aerospace programs to a strong management approach to emphasize short-term planning, risk mitigation, and adaptability to changes as well as close collaboration with the customers.

Companies, institutions, military programs, and successful businesses have been going through their own evolution. Customers are informed and won't hesitate to abandon a solution if they perceive a lack of quality. Expectations arise from convenience and functionality

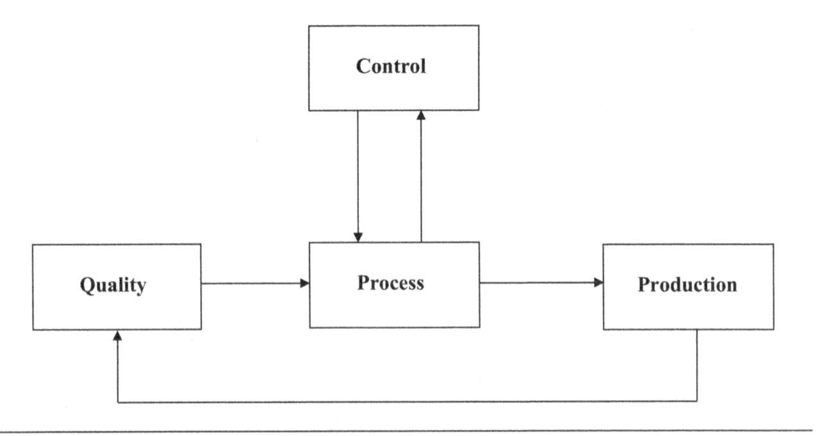

Figure 1.4 Agile management model.

to stand head and shoulders for organizations to deliver products and services that are immediately relevant and offer real value. The window of opportunity grows to innovate, develop, and roll out high-quality products faster than ever before.

Moving ahead with Lean and Agile navigates changes that call for organizations to adopt a more agile stance, one that balances speed with precision. The right solution will accelerate and connect strategy to delivery of products to customers and produce effective deliveries.

A natural result of deploying Lean and Agile is an improved collaboration at the team level and should seek out the boundaries across tools, teams, and workers with real-time information that can help them excel.

1.12 Proactive Approach to Quality Assurance

Quality Assurance consultants must support by encouraging companies, institutions, military and aerospace programs, and successful businesses to be cooperative, and a proactive approach to quality ensures process compliance through evaluation and management participation.

Compliance verification is performed using quality evaluations, assessments, reviews, or appraisals. A Quality Assurance consultant witnesses/monitors activities in accordance with the project-level reviews and meetings. Review Table 1.1.

Quality Assurance consists of the means of monitoring many engineering processes and methods used to ensure quality. The method by

Table 1.1 Quality Assurance Activities

INPUT (PRODUCT/SERVICE)	OUTPUT (PRODUCT/SERVICE)
Proposals	Noncompliant audit reports
Contracts	Evaluation records
Companies/institutions	Activity records
Military programs	
Work product reviews	
Verification of records	
Verification plans	
Verification procedures	
Verification results	

which this is accomplished are varied and may include ensuring conformance to one or more standards. Quality Assurance encompasses the entire process, which includes processes such as requirements definition, control, reviews, configuration management, testing, and product integration.

Quality Assurance is organized into goals, commitments, abilities, activities, measurements, verifications, and validation in supporting processes that has to provide the independent quality in which all the work products, activities, and processes comply with predefined plans and procedures.

1.13 Quality Assurance First

There are many effective methods for audits and evaluations, but the primary method is to first ensure Quality Assurance and to always remember that other methods are secondary. This has always been talked about in all the years of working in software engineering for Quality Assurance for many military and aerospace programs for The Boeing Company.

Software engineering for quality has a major task in all reviews, audits, and evaluations. I would like to see quality teams step up, take charge, and be responsible in preparing and show performance. What it takes is being involved every day of the weeks, months, and years. I know that a Quality Assurance role is to be independent, but get ingrained, and you will see confidence and support from all personnel you work with.

I have attended many software meetings to verify software builds, software loading in software integration labs, and production environments. The military and aerospace program management team, systems engineering personnel, software teams, testing team, and software leads respected me due to my involvement to show Quality Assurance first. It was time for the software engineering quality personnel to make this change and an understanding to management and team members.

For years I have witnessed and performed software Quality Assurance activities supporting the military and aerospace programs. As a software engineering quality representative, I performed Quality Assurance software reviews, audits, and evaluation daily, involving all managers and test team members. In performing engineering reviews and audit activities along with effective evaluations, you can accomplish high expectations and the assurance that prompt deliveries to the customers show compliance to all requirements.

"Remember Quality Assurance is FIRST"

1.14 High-Performance Work Teams

Working as a software engineer and implementing Quality Assurance tasks and activities in the United States and international countries, I have seen so many issues and problems resolved by the High-Performance Work Team (HPWT). I remember an experience working for a military and space program, where we faced many problems as the software worked correctly inside integration labs.

1.15 Plan Implementation

The upper management sets up a meeting to discuss these issues and problems. I remember going to the meeting that comprises both hardware and software teams. While walking into the meeting, I could see the hardware people on one side and the software people seated on the other side of the table staring at each other. The upper managers who came to the meeting were frustrated with the activities going on inside integration labs. Questions were asked by managers that we need to fix the issues and problems. Where are all these issues and problems coming from? The hardware people pointed at the software people, saying they are the problem. The software people pointed at the hardware people and saying they are the problem. I was sitting in the back of the room, and I could not believe what was going on. One manager said: Where is Boyd Summers? The manager knew I was a Quality Assurance employee. I raised my hand, and the manager pointed me to fix these issues and problems and complete the task in 1 week. I said that my tasks were to normally look at Quality Assurance processes and perform audits. The manager told me that he knew what I did and that I needed to look into the ongoing activities in the integration lab and see what the issues and problems were and report back to the management team. I said the work would be taken care of, and the results would be reported in a week. That is the day I decided to put together an effective HPWT.

1.16 Tasks and Results

HPWTs are required to meet daily and report the problems that they might face each day inside the integration labs. The task I took on was to manage the software development process and work with the software engineers and witness the processes they go through.

Being a software engineer and programmer, there was a process that I always followed called Peer Review. In working with the software programming team, I could see that Peer Review were not being performed. Peer Review is a process that allows other software programmers to review and ensure that coding and early testing are working. Many of the software programmers felt that coding and testing were not needed to be looked at or reviewed by others. I set up a meeting with the software engineering team to make sure Peer Reviews are being conducted with the other software engineers. I emphasized that Peer Review is an important part of verification and will be a proven mechanism for effective defect removal. When conducting informal or formal peer reviews, ensure that guidelines are understood by the team members from the start to the finish. Team assignment responsibilities included the data collected for each peer review and the tools used to establish, collect, and require data.

After 2 days, Peer Reviews were performed, and the software was ready to be sent to the integration lab for testing. Review Table 1.2.

Once the software was tested and provided to the integration labs, I had comments and documentation from the HPWT to be used, witnessed the use of the latest software to be installed, and worked with system and software integration. After 2 h of witness, verification, and validation, the integration was a success and was ready for delivery to the customers. The hardware was integrated and working, build

Table 1.2 Peer Review Method

INPUT	INPUT TO FUNCTION	PROCESS STEPS	OUTPUT REQUIRED
List of products being reviewed	Management and technical staff	Plan for peer reviews	Peer review plan
Review package	Quality and team members	Prepare the work products for the review	Review package
Review package	Review participants	Conduct the review	Package identified with errors, defects. and action items
Review package with identified errors, defects, and action items	Review participants	Complete the review	Completed review package and accepted work product ready for build and release

of the software was verified, and testing was performed. As a Quality Assurance representative, I was able to verify all activities performed.

The HPWT was happy and meeting with the upper management to provide the systems and software integration results, showing that the Quality Assurance (QA) buy offs were to be ready for delivery.

The upper management was happy and pleased with our team's performance. Also, the team became good friends with the hardware and software members. All Quality Assurance representatives provided assistance and helped all managers to be glad and successful; however, they were required to understand the roles and responsibilities of quality engineering. The Quality Assurance factors are essential and important to understand. Moreover, it is important for a Quality Assurance representative to fix issues and problems to show that military programs become successful and have all tasks performed in our day-to-day work to be successful.

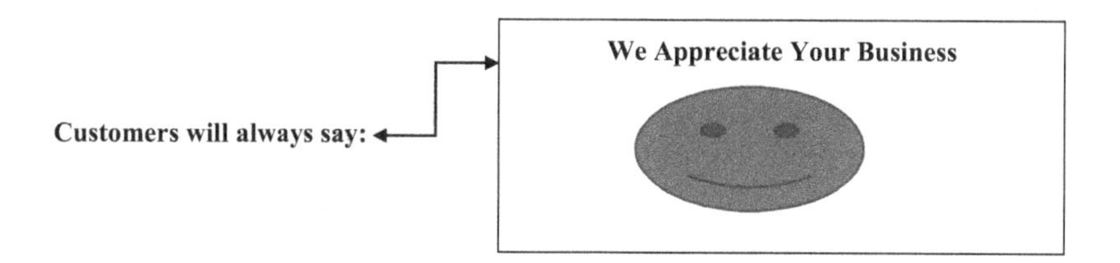

Solving and fixing issues and problems show that improvements can become successful if the following model for process improvements applies as shown in Figure 1.5.

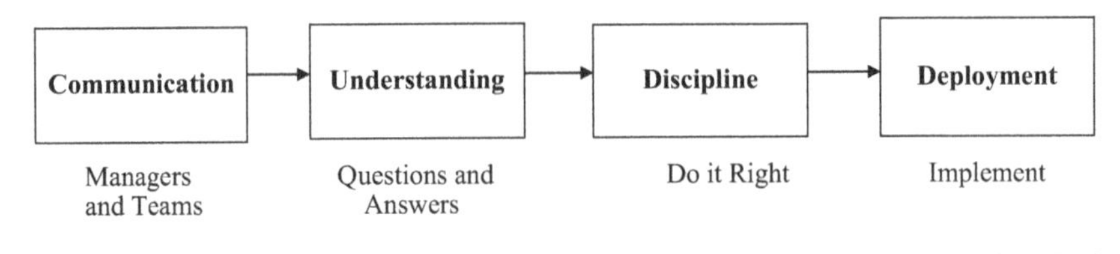

Figure 1.5 Process improvements.

1.17 Quality Management Systems

The term Quality Management System (QMS) is a collection of data process from companies, institutions, military and aerospace programs, and successful businesses that focused on consistently meeting customer requirements. It is expressed as the organizational goals and aspirations, policies, processes, documented information, and resources needed to implement and ensure that processes are followed. QMS emphasized accomplished outcomes of product production line by using statistics and product development. Labor inputs are typically the most important cost inputs to focus on team member cooperation, especially the early signaling of issues and problems via continuous improvement. In this decade, QMS has tended to converge with sustainability and transparency initiatives and customer satisfaction perceived by quality and all developed factors. The acronym "QMS" was coined in 1991 by Ken Croucher, a British management consultant working on designing and implementing a generic model of a QMS within the IT industry.

The QMS is a requirement to have processes documented and execute them with knowledgeable people and teams. At times, metrics are reviewed and monitored to ensure processes are showing improvement. Later, I will have a chapter that defines and talks about metrics and has a good understanding of the importance of metrics and how they come into play with Quality Assurance.

All customer focus should be on QMS and should follow the provided framework: say what you do, do what you say, prove it, and show improvement. The Standards for QMS is AS9100 AS9100C, AS9100D, SAE AS9110, and ISO 9001 and is the model for

- Quality Assurance and Control
- Design and Development

- Companies, Institutions, Military and Aerospace programs, and successful Businesses
- Production and Delivery Results
- Policy and Compliances

1.18 Quality Assurance Teams

The Quality Assurance team or organizations provide for effective product evaluation processes and specific Quality Assurance for effective business methods. The Quality Assurance team does ensure compliance to required standards and control of all developed work products and changes.

It is important that quality attributes become the responsibility of everyone supporting development environments inside business companies, military and aerospace programs, and projects.

The management of Quality Assurance for all activities is summarized in reviews, changing control, customer audits, and compliance to contracts, standards, verification, and validation.

Many managers can be very uncomfortable when presenting what their company can provide to customers. Give them a chance and let them relax. Be positive, some companies will not select you as a subcontractor but move onto the next customer. The learning process will benefit upcoming businesses for understanding what customers will want to hear. Again, *Be Positive* in all Quality Assurance tasks and activities.

1.19 Customer Satisfaction

The concept of ensuring to a business customer that effective work methods have been compliant does meet concrete expectations for customer satisfaction. Many mistakes are often made when business companies work with poorly defined required expectations and requirements from the start. Poor execution will compromise on the Quality Assurance of important and required methods to produce and deliver value. All managers need to ensure that the expectations from the work employees and customers are understood and that these expectations have no surprises to wreak havoc.

When poor Quality Assurance methods are not effective; that can lead to major problems with management, team members, employees and customers.

Boyd L. Summers

Further Reading

AS9100 Standard Quality Management System Requirement Guidelines for the application-Part 2-1997 SAE AS standard, Quality maintenance system requirements for Aviation, Space and Defense Organizations-2009-01.

Lines, M., & Ambler, S. W. 2012. "*Disciplined Agile Delivery: A Practitioner's Guide to Agile Software Delivery in the Enterprise*", IBM Press. 544 pp. ISBN 978-0-13-281013-5.

Rose, K. H. 2005. "*Project Quality Management: Why, What and How*", Fort Lauderdale, Florida: J. Ross Publishing. p. 41. ISBN 1-932159-48-7.

2
Quality Assurance Methods

The purpose of Quality Assurance Methods is to provide a common operating framework in which best practices, process improvements, and cost avoidance activities can be shared and Quality Assurance responsibilities assigned.

The expected results from Quality Assurance Methods are shared for best practices such as improvement process execution and reduction of operational costs. To be internally consistent, the definition for Quality Assurance will be used throughout this book. This definition does not in any way prevent the applicabilities such as quality engineering, quality specialists, and engineering Quality Assurance.

2.1 Quality Assurance Methods

Quality Assurance Methods are a way to ensure that productive capacity meets its intended end results, such as customer requirements, planned production levels, reliability, risk, and direct economic cost. The term is used for companies, institutions, military programs, and successful businesses as far reaching as production and technology. Quality Assurance Methods provide assurance related to productivity and quality management due to organizational and human factors as well as technical aspects.

Companies, institutions, military programs, and businesses will always be successful once personnel encouraged to be cooperative have a proactive approach to the importance of Quality Assurance Methods. Quality Assurance Methods ensure process compliance with the standards and contractual agreements through evaluations, audits, and effective meetings along with management participation. The results are always reported to the management.

For the purpose of clarity and specific direction, site instructions referenced will be used to conduct Software Quality Assurance (SQA) activities.

A Quality Assurance representative should always be responsible for assuring compliance with the contract and adherence to compliance, and effective methods always fix issues that would occur on a daily basis. Each Quality Assurance representative will be trained on methods and effective tools needed to perform evaluations and audits. Compliance verification is performed using process evaluations, assessments, reviews, and appraisals. A Quality Assurance representative witnesses/monitors day-to-day activities and is also an active participant in business and project-level reviews and meetings.

This independence provides an important and needed strength to the Quality Assurance organization, allowing audits and reviews to provide management with many objective insights to business companies, institutions, and military programs. The traditional Quality Assurance methods involve the identification of design/development to provide quality control. Selected work products and descriptions maintain traceability of those life cycles and are key points throughout all Quality Assurance Methods.

2.2 Functional Quality Assurance Tasks

Functional Quality Assurance tasks, along with coordination and implementation of business companies, institutions, and military programs, will be performed. Responsible functional Quality Assurance representatives will review and approve any tailoring or waivers of quality processes implemented and used. Support from personnel can also provide initiatives and tasks for common software process implementation, measurement, and improvement.

2.3 Perform Effective Process Evaluations

Effective process evaluations are used to provide business companies and military and aerospace programs with the process steps necessary to conduct evaluations of all developed and released products. Process evaluations ensure that all products meet its specified requirements and are accomplished through various activities.

Required actions for process evaluations are contained in the following descriptions:

- Process evaluations criteria will be defined from previous audit history, organization procedures, development plans, or other documented requirements.
- Process evaluations will be scheduled based on schedules and activities.
- Results will be recorded, including any noncompliances or observations.
- A process evaluation report will be generated that contains at a minimum: area audited, scope purpose, completed checklists, criteria, participants, results, noncompliances/observations, and lessons learned for future improvement.
- The process evaluation reports will be reviewed and approved by management or designee and distributed to relevant stakeholders.
- Actual measurement data generated during the conduct of this method will be collected, and noncompliances will be addressed.
- Based on contract requirements, business unit, requirements, software development milestones, and project-level plans, ensure that software products that require evaluations are identified.
- Process evaluation criteria and plans established based on contract requirements, business unit, requirements, delivery milestones, and level plans.
- Process evaluation records document and identify the products under review, evaluation criteria, participants, results, and findings. Identify any lessons learned and propose process improvements that will improve per evaluation process.
- Resolution of findings elevates unresolved issues as needed. This will occur before customer delivery.

2.4 Consistent Quality Assurance Process and Reviews

Consistent Quality Assurance processes and continuous reviews provide effective process execution and will show proof of compliance. These stages empower employees to always understand and use the right Quality Assurance processes as shown in Table 2.1.

Table 2.1 Quality Assurance Processes

CONCERNS			
No one is perfect, but can work to ensure Quality Assurance is implemented correctly	Problems begin when you allow and show disappointment	Tell management and employees that they must step it up	Pride is good and show positive tasks during the implementation of Quality Assurance

To improve Quality Assurance processes, definition and improvement allow access to content (e.g., plans, documentation, contracts, requirements) and ensure compliance that will make a strong and workable management team. As a Quality Assurance auditor, you must show execution to manage results for all assets. The tailoring for Quality Assurance processes is as follows:

- Guided processes are tailored for consistent reviews.
- Process compliance for work products is reported to the management.

2.5 Documentation and Required Data Control

All documentation and required data control are processes by which entities review the quality of all factors involved in production, as ISO 9000 defines data control as part of quality management focused on fulfilling quality requirements as an emphasis on three aspects enshrined in standards such as ISO 9001:

- Elements such as controls, management, defined, and well-managed processes for performance and identification of records.
- Competence such as knowledge, skills, experience, and qualifications.
- Personnel, integrity, confidence, motivation, team spirit, and quality relationships.
- Inspection is a major component of reviewing documentation and required data control, where a physical product is examined visually and the end results are analyzed.
- Product inspectors provide lists and descriptions of unacceptable product defects.
- The Quality Assurance of the outputs is at risk if any of these three aspects are deficient in any way.

Companies, institutions, military programs, and businesses establish all documents, procedures, and work instructions and any other data that show evidence that all employees use to provide Quality Assurance aspects. All procedures provide explanations and how standards are to be implemented and applied. Data control of technical manuals from employees and management must be controlled. The management team will provide information for deciding what is controlled. All documents must have identification of changes and at times highlight those changes. Documents should always be available to employees who need and use them. A system to mark outdated documents is maintained for later use or for legal purposes.

2.6 Basic Standards

The element of basic standards requires identification of plans and procedures for production and service that can affect Quality Assurance processes. There are elements that should always be addressed. These elements are as follows:

- All plans and documents show how work is done.
- Effective tools for handling work used in a working environment.
- Compliance to monitor and control work products.
- Approval of Quality Assurance processes.

Quality Assurance auditors should always assess the operations where all the work is done. The auditors need to talk to personnel and ensure they have the training and experience and the knowledge of process control for all data documentation. The auditors will interview personnel and ask about the workmanship activities, specifications, and tie them to records. Education and training of personnel are the required standards and must always be correct. The standards offer ways to address specific processes by continuous monitoring of all processes. All Quality Assurance auditors need to demonstrate the capability to deliver effective and efficient data. There should always be an ongoing training program to stay updated and show improvement to satisfy customer needs.

In basic standards, the needs focus on the measurement of what is believed to be in development programs following the basic standards to approach in economically productive activities that will help companies, institutions, military programs, and businesses. Basic standards carry its own weight in the future and focus on allowing to rise above the development line to meet its basic standards. Programs focus on fairness in terms of measurement and the basic need of importance to be successful.

2.7 Quality Assurance Stability

Without Quality Assurance stability, uncontrolled tasks can occur, and thereby have no recognition issues and concerns that could cause improvement actions. There could be no repeatable Quality Assurance processes to help with process improvement. Process capabilities for companies, institutions, military programs, and successful business abilities need to be performed per plans to ensure that products are compliant for better product work environments and be competitive for release to customers.

Quality Assurance processes are under static control and distributions of products per process performance for continued operations of effective processes performed. Continued operations of Quality Assurance processes will always remain stable to ensure that measured product values are consistent and compliant per measurement capability. It is critical to understand and implement the disciplines during life-cycle tasks and activities before delivery of quality products. Quality Assurance methods and stabilities will benefit current and future companies, institutions, military programs, and businesses.

Further Reading

Cassidy, A., 1998. "*Practical Guide to Information Systems Strategic Planning*", Boca Raton, FL: CRC Press.

Jolly, R., 1976. "The world employment conference: The enthronement of basic needs". *Development Policy Review*, *A9*(2): 31–44. doi:10.1111/j.1467-7679.1976.tb 00338.x.

Managing Quality Across the Enterprise: Enterprise Quality Management Solution for Medical Device Companies. Sparta Systems. 2015-02-02. *The Quality Assurance Journal*, ISSN 1087-8378, John Wiley & Sons.

3

Measure Benefits of Quality Assurance Process Improvement

The major obstacle in measuring the benefits of Quality Assurance Process Improvement for companies, institutions, military programs, and successful businesses is to invest in Quality Assurance and ensure that processes are implemented and followed. This chapter provides a summary of evidence of business benefits for Quality Assurance improvement. Issues are discussed related to measurement of Quality Assurance performance, and an outline is developed for starting compliant processes.

Most companies, institutions, military programs, and successful businesses should always have a quantitative approach to manage staff members and employees for business performance. In the software industry, I have worked daily with measurement processes to evaluate many software engineers. Many of the software engineers do not like having measurement to be presented, but managers like to have measurement capabilities. Measurement of Quality Assurance Process Improvement is necessary to monitor the performance of staff members. It is equally important that measurement becomes common sense and ensures that measurement can improve the results of all companies, institutions, military programs, and successful businesses to address wrong issues. One of the main objectives of Quality Assurance is to improve and review measurement data to achieve statistics and metrics for product production and processes. Measurement performance of Quality Assurance processes provides feedback, which is important and necessary for all members to evaluate employee performance and to ensure improvement. There should always be continuous Quality Assurance process improvement. Refer to Table 3.1.

Table 3.1 Quality Assurance Performance

	PROCESS RELATED		
Number of defects per analysis	Number of process issues and problems	Productivity activities	Quality assurance trends
	PROJECT RELATED		
Development time	Products timed	Project development times	Members and employees productivity

3.1 What Is Needed for Measurement?

When collecting quantitative Quality Assurance measurements for Adversity and Concerns, you must be careful on the decision to be measured. Performance must be in place and care taken when decisions and measurements are to be used for Quality Assurance and all evaluation criteria. Many measurements could cause confusion and distort the overall operations being performed per Table 3.2.

Needs for Quality Assurance measurements for operations and project performance objectives are dictated by operations management policies. Operations management, by definition, focuses on the most effective and efficient ways for creating and delivering effective products that satisfy customer needs and expectations. As such, it ties with effective processes for Quality Assurance and the five performance objectives that give quality acts as one of the five operations/project performance objectives provided by operations management policies as

Table 3.2 Adversity and Concerns

	ADVERSITY ISSUES		
Don't be frustrated doing your job for Quality Assurance	Show leadership advice to management and employees	Always stay positive and not let management and employees see you upset	Show that you are a professional and doing your job well
	CONCERNS		
No one is perfect, but can work to ensure Quality Assurance is implemented correctly	Problems begin when you allow and show disappointment	Tell management and employees that they must step it up	Pride is good and show positive tasks during the implementation of Quality Assurance

- Quality Assurance measurement conforms to accurate specifications.
- Measuring delays between customer requests and product service.
- Dependability on measuring how consistently a product or service can be delivered to meet customer expectations.
- Measuring how quickly companies, institutions, military programs, and businesses can adapt to a variety of changes.
- Cost and measuring the products are required to plan, deliver, and improve the delivery to customers.

Based on these objectives of support, Quality Assurance increases dependability, reduces cost, and increases customer satisfaction.

3.2 Quality Control

For many years, I have been attending and lecturing at many Healthcare conferences in the United States and international countries, and alongside have submitted many articles related to Quality Assurance and the importance of Quality Control. The company is One Medical Issue Clearly Sufficient (OMICS), and the topics relate to Good Management Practice (GMP), General Medical Clinic (GMC), and quality control for medical businesses and institutions. The healthcare in the United States is very similar to that in the international nations. It is important to handle and deal with quality control and setbacks with problems related to healthcare and strategic planning and to ensure healthcare companies will be successful. The best way is to be aware of how we can handle frustrations to carry out my job in Quality Assurance.

Quality control begins with collection and reporting of plans and procedures that are achieved through control of performance. There should be initial control of what can be achieved through specification of services, tasks, and accomplishments. Daily performance must be documented first to ensure continuous production and to ensure that the data is accurate. Daily checks help certify that the process for companies, institutions, military programs, and successful businesses is measured to show compliant and successful tasks and duties.

The quality control efforts depend on the training, professional pride, and importance of a particular project to be compliant. The burden of a manager and team member can be lessened through the implementation of Quality Assurance programs. Through the implementation of established and routine Quality Assurance programs, two primary functions are fulfilled: the determination of quality and the control of quality. By monitoring the accuracy and precision of results, the Quality Assurance program should increase confidence in the reliability of reported results.

A quality control culture greatly depends on the management, team members, and employees to define what behaviors are appropriate for process improvement. The tasks and activities for quality control will determine which behaviors are good and add value towards goals, knowing that it is important because it helps define risks that organizations can take to help manage organizational support.

3.3 Problems with Healthcare Do Need Quality Assurance

The problems with ensuring Quality Assurance will increase healthcare risks and can impact the quality of life. Effective risk assessment methods will enable healthcare businesses and institutions to be more willing to resolve issues and problems that could occur with medical patients. With the rapid changes taking place in healthcare these days and months, business companies need to seek effective processes for Quality Assurance. Improvement can be accomplished to achieve success and excellence in all organizations that can make competitive organizations at all levels.

Healthcare is reviewed into three classifications:

- Clinical peer reviews to provide academic Quality Assurance practices.
- Evaluations of clinical skills needed for both physicians and nurses.
- Reviews and understanding of medical and healthcare journal articles and statements.

Additionally, medical peer reviews can be used in the process of improving Quality Assurance and safety in healthcare. The process of rating clinical behavior or compliance with professional membership standards is necessary and important.

3.4 Control Charts

Control Charts are used as a statistical tool for handling and solving problems to ensure that Quality Assurance process are under control. The Control Charts tool provides graphics that can aid in the study of process improvement that contains measurement, analysis, and control for continuous improvement. The strategy for Quality Assurance will need to rely on all company goals and objectives to provide all projects, programs and company stakeholders to become more effective as a guide for management and employees to become successful.

The complex interrelationship between analytical methods and charts show concentration, detection, and method processes that management of Quality Control is undertaking using a statistical approach to determine whether the results obtained are within an acceptable process and activity.

3.5 Strategic Planning for Quality Assurance

The strategic planning will always examine and show successful planning for the implementation of Quality Assurance. This type of strategic planning improves management and employee services to be analyzed to ensure that Quality Assurance is integrated and show results that will indicate that this type of planning is influential for implementing Quality Assurance to be successful.

3.6 Commitment to Quality Assurance

It is important that companies, institutions, military, and successful businesses show commitment to Quality Assurance and specialize in creation of effective solutions that meet all customer expectations. The goal is to meet the expectations of management and employees, be successful, and have an effective experience to ensure that Quality Assurance works meets all expectations as well.

Every step management and employees take will need to ensure that Quality Assurance is used for healthcare purposes. There should always be importance for Quality Assurance monitored by experienced staff members and employees. With many Quality Assurance experts around the world, make sure support is provided for technical content and tools used. When help is needed or assistance discussions for requirements, customer service and support from Quality Assurance representatives are ready to help and provide solutions for success.

3.7 Software and Quality Assurance Improve Healthcare

Being a software engineer for 30 years, I can see that developed software will improve medical outcomes for healthcare. Results from software are promising and enable management and employees the opportunity for effective programs and increase patient knowledge supporting health and treatment. Focusing on problem-solving using the software can help solve many issues and concerns that will be useful and progressive toward ensuring healthcare in working to support all doctors, nurses, and patients. Quality Assurance process is very important in the medical and healthcare field, because it helps to identify the standards of medical equipment and services. Quality Assurance is particularly applicable throughout the development and introduction of new medicines and medical devices. The Quality Assurance organizations support and promote the quality of research in life sciences through its management and team members.

3.8 Quality Assurance Process Improvement

Quality Assurance Process Improvement encompasses all activities for all companies, institutions, military programs, and successful businesses and can lead to better process for high-quality deliveries to customers in a timely manner. The approach to process improvement for Quality Assurance can be viewed in the following steps:

1. Education and training for management and team members.
2. Selection of process activities for engineering and tool development.
3. Implementation of Quality Assurance Process Improvement plans.
4. Reviews, audits, and evaluations based on the results of plans.
5. Perform assessments of all current process implemented and used.

The approach to emphasize process improvement is to examine well-defined tasks and activities to ensure all processes meet Quality Assurance expectations. The important knowledge is to understand all process workflows and ensure accomplishments are used to support all technology transitions and measure all changes that have been adopted. Education and training will benefit solid processes and practices that can lead to decisions due to Quality Assurance Process Improvement when introduced. A key element is when management has direct contact with all organizations and provides team members and employees the necessary tools adopted for use. Quality topics focus on better communication that will foster best process improvements.

Further Reading

Majcen, N., & Taylor, P. (Editors) 2010. "*Practical Examples on Traceability, Measurement Uncertainty and Validation in Chemistry*", Vol 1. Belgium, JRC. ISBN 978-92-79-12021-3.

Pressman, R. S. 2010. "*Software Engineering, A Practitioner's Approach*", New York, McGraw-Hill. ISBN 978-0-07-337597-7.

Pyzdek, T. 2003. "*Quality Engineering Handbook*", Quality Assurance vs Quality Control – ASQ Journal. Boca Raton, FL, CRC Press. ISBN 0-8247-4614-7.

4 Quality Assurance Performance and Improvement

All companies, institutions, military, and successful businesses need to advance Quality Assurance performance and improvement to ensure processes are followed and are compliant to all needs. A critical part of implementing Quality Assurance performance and improvement enables management, team members, and staff employees to report issues and events that might affect the ability to deliver services of very high standards. This mechanism is called Quality Assurance Performance and Improvement Reporting.

The Quality Performance and Improvement Report is used as the basis for quality efforts—to look constructively at how services and workplace can be safe and more secure, efficient, and consistent. Remember that the report is not about finger pointing or apportioning blame; instead, it helps to be more mindful about what is done and how to accomplish and identify aspects that might need extra support to achieve the highest levels of Quality Assurance.

4.1 Progress

Progress has many stages that management and employees need to follow and know their roles and responsibilities to ensure that Quality Assurance performance and improvement is working. In my experience, better is better and can sometimes cause an overload on performance and improvement, but always remember to focus on two topics.

4.2 Preparation for Quality Assurance Performance

Guidelines for the preparation and accomplishment of Quality Assurance performance: One of the guidelines and values from performance is the ability to prepare daily ideas or concepts that provide the objectives to be accomplished.

The main ideas relate to making correct statements to identify subjects to be covered and to ensure costs and schedule for companies, institutions, military, and successful businesses. For example,

- Capability of performance is required.
- The cost should be reasonable for future values.
- Important schedules can be maintained and effective with management support.
- A Quality Assurance performance indicator is a type of performance measurement. Performance evaluates the success of companies, institutions, military programs, and successful businesses, in particular, activities such as projects, programs, products, plans, and other initiatives.
- Often, success is simply the levels of operational goals and is defined in terms of making progress toward strategic goals. Accordingly, choosing the right performance relies on a good understanding of what is important for management and team members for measuring performance used for assigned tasks and actions.
- Since there is a need to understand what is important for various techniques to be assessed, it is important to identify activities that are associated with the selection of performance indicators. These assessments often lead to the identification of potential improvements, so performance indicators are routinely associated with "performance improvement" initiatives. A very common way to choose performance is to apply a management framework.

The Quality Assurance performance and improvement will establish standards relating to Quality Assurance, and performance improvement with respect to facilities provides technical assistance to companies, institutions, military, and successful businesses to meet such standards. Best practices include how to coordinate the implementation of plans and procedures with quality assessment and assurance activities.

4.3 Preparation of Quality Assurance Improvement

Roles and missions describe the scope of Quality Assurance Improvement capabilities. Key results for improvement relate to management and employees having successful objectives and show that the results are achieved. Action plans provide the sequence of actions to be implemented in order to have improvement for all companies, institutions, military, and successful businesses to reach successful accomplishments. Management needs to have communication to serve as a process to lead and show employees that are involved in decision making and show commitment to carry out all decisions. Quality Assurance improvement can be a simple process due to having common sense and a logical approach to improve principles and techniques for management responsibilities.

4.4 Evaluation for Quality Assurance Processes

To provide evaluation for Quality Assurance processes is to look at the time spent in each development phase. When management, employees, and Quality Assurance personnel spend more than 50% of detailed time, experience will show that evaluations are reasonably thorough.

To achieve high Quality Assurance tasks, always spend time knowing that review times, audits, and inspections measure the importance of performance and improvement. I suggest using the following guidelines:

- Requirement and contract reviews used for inspections.
- High-level reviews and inspections.
- Detailed data reviews and audits.

It is effective to use checklists in which issues or concerns related to management and employee discrepancies are updated frequently. When the quality data reflects problems, review with deep content to find the cause that indicates many problems. After the completion of Quality Assurance evaluation cycles, experience will come into play for adjustments being made and accomplished. A very common approach is to document these adjustments in a Quality Assurance notebook and, if needed, document reports. Strive to improve process and Quality Assurance to be successful during business activities performed by management and employees.

4.5 Quality Assurance Excellence

Management and employee teams can help in the drive for Quality Assurance excellence and always show the start of the importance of Quality Assurance for all companies, institutions, military, and successful businesses. Make discipline a habit to ensure that knowledge and skills achieve all expectations required by management. Before habits happen, there should be three topics that are needed:

- The knowledge of why habits are important.
- Have the skills needed to accomplish lessons learned.
- Have all disciplines to show desire is the key element.

Motivate desire to be consistent for striving for excellence in all companies, institutions, military, and successful businesses. A desire can be something that is always a method that all professionals should need, but having management and employees included is where working together with defined processes and goals can help in working together to meet all expectations.

Note: When a manager or employee does not follow processes, there should be pressure applied to make sure conformance is known and important. Motivation and maintaining disciplines help to establish processes, plans, and goals and monitoring of these actions on a daily basis can be applied to be effective for excellence.

Achieving excellence can be a struggle, since the world is in a change, and facts are astounding each day. Human capabilities increase every year, and the key element is that humans excel. When people work to excel, they often find that they can work better, strive to improve, and have better management and employees to be responsible on how to handle problems. That is when Quality Assurance can help management and employees to work things out, step up to all challenges, have a convincing Quality Assurance plan, and have management agree to what is proposed. Employees must care about the quality of all products produced and be consistent with all effective methods to reach that capability. Excellence can start with employees because they know that the world is changing and that the focus on improvement is important to be successful.

4.6 Building and Maintaining Quality Assurance Teams

Building and maintaining the Quality Assurance teams will have the obligation to have a coherent and effective team for all companies, institutions, military, and successful businesses.

Quality Assurance teams need to meet and discuss the following:

- Are clearly and defined goals important to management and employees?
- Are plans in place to achieve these goals?
- Make sure that everyone is involved in making plans.
- Make all Quality Assurance team members show commitment to meet all developed plans.
- Having weekly meetings to review status, plans, issues, and accomplishments.

Comment: If all Quality Assurance team members are successful, you will succeed.

4.7 Quality Assurance Problem-Solving

Quality Assurance problems occur when the teams ignore the peer pressure from management on topics that could cause threats and discouragement. In addressing these Quality Assurance problems, the team should review these topics and make positive help to resolve the topic presented by management. If problems are discussed and responded quickly, it is important for the team to call for help to resolve difficult problems. Never feel that you are alone and insist you will fix problems with others and meet specific goals. Support from other team members is critical and will provide you time in helping management and employees resolve problems provided to you and team members. This is the key in providing effective support and show that team members believe in their abilities to solve Quality Assurance problems and do superior work. Remember that team members who believe in themselves will often perform beyond what can be possible for all problems, issues, and concerns. Management and employees become successful knowing that the Quality Assurance Team can help to resolve all problems, issues and concerns with which they are presented.

4.8 Common Quality Assurance Working Framework

The importance of having a common Quality Assurance Working framework is important, where the Quality Assurance team goals could be challenged and the path to achieving goals should always be clear. All tasks need to be achievable, and understanding the roles, responsibilities, and how to accomplish the team goals is needed for success. The tasks needed for common Quality Assurance working framework are as follows:

- What are the tasks to be performed and accomplished?
- Provide a timeline for all tasks to be completed.
- Assign team members to achieve completion of required tasks.

All teams need to be cohesive and have challenging goals, performance feedback, and agreements for the framework processes to be defined for all activities for management and employee accomplishments.

The Quality Assurance Working Framework is intended to improve the quality of general practices and is part of an effort to solve issues and numerous problems. Participation is required for most activities under the contracts and can make a difference when daily processes are implemented for Quality. The criteria for Quality Assurance Framework activities are designed around best practice and can have a number of accomplishments allocated for achievement.

In many companies, institutions, military, and successful businesses, the value of accomplishments is always measured for expected goals. The working framework is reviewed, audited, and evaluated. Quality Assurance representatives ensure that processes, plans, and procedures are working to achieve expectations.

4.9 Quality Assurance Impacts Business

It is sometimes easy to ensure that Quality Assurance impacts business when working and meeting program and project deadlines. Quality Assurance representatives can support and understand turbulent outcomes that can arise if required standards are not enforced.

It is clear that businesses fail when insufficient companies, institutions, and military operations fail when insufficient policies are not followed. Management, team members and employees have opportunities to resolve situations to strong understanding, make changes to improve processes, implement change and have open communication for problem-solving.

Here are five actions that can be used:

- Develop leadership to champion Quality Assurance practices.
- It is important that everyone knows the guidelines and expectations.
- Communication is important as a top priority.
- Gain feedback from employees bringing up concerns.
- It is essential to provide strong understanding processes required.

4.10 Ways to Improve Quality Assurance

It takes courage to make clear-cut objectives to improve Quality Assurance. It is worth to create an effective understanding for management, team members, and employees for competing interests and to maintain focus on programs. Always help everyone to stay on track with these six tips to move faster and more efficiently:

- **Set a Schedule**: Go through your task list and review a huge list of awesome ideas for positive actions and scheduling.
- **Invite Helpers to Help**: Enter your tasks into your programs and project management and assign helpers to yourself on your team and make sure every step is accounted for to limit the possibility of delays.
- **Make Your Meetings Count**: All meetings are important work that could be accomplished and include updating your program and project management systems on a regular basis.
- **Communicate Clearly**: It is important to map out the requirements, expectations, and tasks for your team constantly coming to you with more questions.
- **Focus**: If you want to move your programs and projects forward, focus on one project at a time and be flexible if you have dependencies that are waiting to get finished.

Table 4.1 Definitions of Quality

Quality Assessment – an evaluation of a process to determine whether a defined standard of quality is being achieved.
Quality Assurance – the organization structure, processes, plans, and procedures designed to ensure that all practices are consistently applied.
Quality Improvement and Process Performance Improvement – an ongoing interdisciplinary process that is designed to improve the delivery of services and outcomes.
Quality Process – empowers management, team members, and employees to develop level and effective solutions for process development.

- **Goals**: Know what all goals need to be accomplished and move your team toward getting all programs and projects finished.

4.11 Summary

This chapter provides an overview and covers what Quality Assurance teams are and how to resolve problems. When the conducted tasks and activities fail, it is usually teamwork problem and not in-depth technical problems. In building and maintaining the issues of the Quality Assurance team, concerns and problems put pressure knowing that the plans and schedules are aggressive and need to be completed and approved by management. For a Quality Assurance Team to be successful, tasks must be understood, they must have control, and team members should be encouraged to make effective plans. The enjoyment of being successful sustains the enthusiasm and the importance to have management, team members, and employees to work processes and ensure that all processes are followed as in Table 4.1 to show Definitions of Quality.

Further Reading

Berghhoefer, C. 2018. "*Safety Tips that Impact Businesses*", Environment, Health and Safety, Randstad, USA.

Chemuturi, M. 2010. "*Software Quality Assurance: Best Practices, Tools and Techniques for Software Developers*", Plantation, FL, J. Ross Publishing. ISBN 978-1-60427-032-7.

Rubertino, F. 2014. "*Quality Assurance and Performance Improvement*", Danvers, MA, HCPro. ISBN 978-1-61569-357-3.

Russel, J. P. 2003. "*Continual Improvement Auditing*". December 2017. Danvers, MA, HCPro. ISBN 978-1-61569-357-3. http://www.**jp-russell**.com/pdfs/CI-paper-fyi.pdf

5
Answer Questions from Management and Employees

Management and employees who work for companies, institutions, military, and successful businesses always ask questions that need to be answered relating to the purpose of Quality Assurance techniques. As a Quality Assurance representative, you will need to provide answers that meet plans and how the Quality Assurance processes can be effective for all tasks and activities being performed.

5.1 Techniques Required

The ability to answer many questions from management and employees depends on the effective answers from the Quality Assurance representative that must handle questions and answers in meetings, so all management and employees should understand the answers provided. Any questions asked must be answered and should be meaningful to the questioner. It is important that handling questions can be an effective method and meaningful to all questions asked.

5.2 Conduct Questions and Answers

A Quality Assurance representative has the capability to conduct questions and answers from management and employees on a consistent basis. A positive attitude is an important consideration when someone trying to put Quality Assurance representatives will be be defensive. Approach to questions with the idea of answers to those questions implies that they are interested in gaining more and more information that can be provided by a Quality Assurance representative. In fact, this attitude can be completely disarming to anyone who will try and take you down, but always answer questions in a manner that reflects a positive attitude.

Whenever there is a question and answer period, people run at a risk. Quality Assurance representatives must develop skills for dealing with difficult questions that could not be answered in a short time period from management and employees. If there is not a good answer to questions, refer to someone who can provide answers and work the questions later. If time is needed, always evaluate the questions and collect thoughts from others that could be helpful to answer in a positive way. Try one of the following responses:

- Repeating the question would help so that there can be a better understanding.
- Tell that person you provided a very good question and how do you feel about it?
- Think about the question for a few moments.
- Ask how others feel about the question asked.
- Write the question on a chalkboard and provide more thinking time.

More could be written on most critical questions; however, principles and techniques increase success to provide effective answers. Managers play a key role in establishing clear lines of communication within the organizations. Management functions include planning, organization, staffing, leading, and overseeing functions inside and outside companies, institutions, military, and successful businesses and can use networking. All of these functions require communication or work will not get done. By going further to communicate more effectively with employees, you can achieve a more efficient, productive, and satisfying work environment.

Managers must find ways to solicit feedback from team members and employees. This facilitates two-way communication and lets everyone know that their ideas are important. Managers can gain new ideas, as well as insight into jobs. For example, while one employee may be responding to incoming complaints per company policies, another employee may actually have a solution to the problem causing the complaints. A feedback system ensures that valuable employee suggestions are received and implemented.

5.3 Be Aware of Management Mistakes

It is important to ensure that there are zero defects in management decisions and implementing changes. If companies and organizations implement changes, there should always be those changes to be made perfect and understood by all employees. There can be mistakes, so when employees see that management does something wrong, remember it might look right or correct if you would know more.

Management mistakes could look like a wrong move, but could be a simple move to what is required and to be started and completed. The employee's miss of changes could be wrong and the activities compelled for all actions performed by their employees. The term "trade-offs" can be easily looked at as a management mistake. It makes sense that management is trying to do their best.

5.4 Strength Should Never Show Weakness

All work habits and abilities that serve management and employees should always show strength and never show weakness in work-related tasks and activities. Even if job titles and duties remain the same, a situation will call something new from management and employees to work.

Always be sure to shift management and employee job priorities to match changes and always show value and efficiency. Management and employees should adjust their approach to fit and be busy developing new skills to show effective competencies. Management and employees show always retool themselves to ensure employment for years ahead. Be awake, alert, perform, and refocus when required. All jobs should be examined to identify the critical factors that are important for the job's success. It is important to show strength and never weakness and always focus on what is best for companies and organizations.

5.5 Quality Assurance Processes Support Employees

Quality Assurance processes should always support employees that catch a lot of criticism during management changes. Employees need Quality Assurance processes and require help and support from

Quality Assurance representatives. At times, employees will know it is hard to be consistent, and the same thing is true in management directions when things are causing employees a consistent pressure to ensure all Quality Assurance processes are being followed and implemented correctly.

Employees of companies and organizations have the chance to show what they are to do and be loyal and committed to make a change because there's always a need for good employees to protect the companies and organizations they work for.

5.6 Relationship between Management and Employees

The relationship between management and employees is sometimes fragile. The employees may feel pressured to keep their jobs ensuring that everything is done per management directions and wonders if working to the best. If management disregards the concerns of employees, that could lead to a number of important and sometimes major problems. Workplace problem is also a risk to companies, institutions, military, and successful businesses if employees become disgruntled.

Management that doesn't foster a productive relationship between employees and also may develop a poor reputation in the industry. There are solutions that can be taken toward improving the management and employee relationship to facilitate a highly productive and happy workplace. Management can implement employee recognition or reward programs to thank workers for a job well done. Meetings and regular communications between employees and managers are also important, especially if workers feel disconnected from management.

5.7 Quality Assurance Interviews

To build a career in quality it is important for Quality Assurance interviews, questions and answers for companies, institutions, military and businesses working together and become better. It helps you to save time and prepare well for the job interviews related to plans and processes. As a Quality Assurance representative, you have to ensure that the quality of products, services, and processes are properly maintained regularly so as to meet the customer's requirements.

During all interviews, questions should always be asked related to what Quality Management Plan (QMP) expectations are? The answer should be that a QMP documents a management system about all work activities to be performed. The QMP is a document that describes the quality systems in terms of the organizational structure, functional responsibilities of management, team member lines of authority, and required interfaces with those planning, implementing, and assessing all environment-related activities conducted. The benefits of QMP are as follows:

- Improvement in internal quality activities.
- Improvement in external quality (customer satisfaction, conforming products).
- Improvement in production reliability and expectations.
- Improvement in time performance per required processes.

The Quality Principles need to be documented in QMPs to show leadership, process approach, continual improvement, decision making, and relationship with suppliers and customers. Product quality means we concentrate always on quality, but in case of process quality we must set the process parameter.

When quality is the end point, then "quality management" is the approach and process for getting there, and documenting in the QMP provided the needs to consider the key principles that are central to the topic of quality. If we are concerned with providing values, we must consider how we can improve customer value. There are a number of principles that are central to the practice of quality management and must be documented in the QMP.

5.8 Summary

The reality of answering questions from management and employees at times could never be the same for all companies and organizations. The only thing Quality Assurance processes is to make sure that management and employees questions are answered daily by Quality Assurance representatives. When questions are answered by Quality

Assurance representatives always provide a presentation that is meaningful, accomplishment and having answers to all questions.

Further Reading

Johnson, K. S. 2018. "*Effective Communication Between Management & Employees*".

6 Manage Risk Management

Risk management is always managed by Quality Assurance and management of all companies, institutions, military, and successful businesses. Many risks related to exposure can be uncertain, and potentially bad consequences can have a negativity effect for projects that show many uncertainties.

Project management is a form of risk management in that where there can be uncertainty that could have effects. The purpose of understanding risk management is to show negative factors and will point out positive aspects of risk management from time to time.

6.1 Types of Risk Management Projects

There can be many risk management projects that are considered internal and are associated with technical risks in companies, institutions, military, and successful businesses. The degree of risk management can be internal or external and will determine attention that can affect many employees that have concerns because risks can have greater issues. Quality Assurance representatives and managers must look at both internal and external risks to ensure that there are no impacts to projects and important budgets used by many companies, institutions, military, and successful businesses. Risk management is a factor that should always be looked at related to budget and how likely that task impacts:

- Make sure that risk management does not impact schedules and budget.
- Add more resources to fix risks related to budget.
- Risk management could be a hazard to budget but can be fixed but for usage.

Quality Assurance representatives that are assessing risks need to apply or assign a completion date for managers to resolve many issues related to budget concerns that could face a quality risk. It may not be easy to determine risks to the quality aspects that could have schedule problems and costs impacted. Dealing with risks must have Quality Assurance representatives' needs to analyze each risk associated with all project and budget concerns.

6.2 Effective Planning for Risk Management

The more critical a risk can be, provide a backup plan. Establish an effective plan for risk management for managers and employees and also provide training to enable team members to keep the plan and activities moving. Risk management plans are not simple far-reaching goals and are a way of fixing anything related to project and budget concerns.

6.3 Risk Mitigation and Monitoring

Risk mitigation and monitoring is built into all plans prepared by Quality Assurance representatives, taking precautions when making important and effective decisions.

Make sure that risk mitigation is effective for

- Tasks not affecting mitigation in numerous areas.
- Projects that are still cost-effective.
- Producing quality results.

Risk monitoring can be effective when Quality Assurance representatives have an adequate system of tracking many risks occurring based on evaluations at various times related to projects. A long-lasting or difficult project will require more monitoring. If unable to meet the necessary time commitment to changes, risk monitoring will reevaluate every time a project is delayed.

Risk onitoring can accept changes from the original plan. Costs and labor hours and other factors will not be the same as developed in the original plan. As Quality Assurance representatives monitor risks to ensure variance is acceptable and show higher risks for future problems and determine carefully and act or not. When there is date for when deliverables are needed, determine delays to time and costs. Remember risk monitoring will keep situations that can be prepared, and infected risks could present many surprises. Basic rules for risk monitoring are as follows:

- Make sure expectations from management and employees are assigned.
- Make sure all reports are understandable for management and employees.
- Monitor to ensure that there is control related to projects.
- Monitor tracking and gain insight to all information.
- Make sure monitoring is not tough for management and employees.
- Keep status reports short and to the point for understanding.

6.4 Risk Management Activities

All risk management activities need to be done on time and correctly. It is always good to have data and encourage management and employees to discuss issues, matters, and problems during meeting that are to be setup by Quality Assurance representatives. Don't ever take sides but work as a partnership and what is expected for everyone. All people in the meetings need to know what is expected to address risk management issues and encourage communication that will show commitment to all issues and problems discussed.

Evaluating external risks will follow logic that is used when trends, conditions, and other factors could affect many activities. When companies, institutions, military, and successful businesses begin to struggle, Quality Assurance representatives need to evaluate external risk issues that could affect the possibility of being successful in solving risk management concerns and problems. At times, assessing risk management activities is a place and time to make decisions on how to handle risk issues effectively and will benefit in having a very good understanding on how to fix problems. Management and employees during meetings will identify potential risks that possibility has never been aware of and knowing that effective discussions will give everyone a chance to resolve and contribute to provide information and a better course of actions.

6.5 Risk Management for Quality Assurance Process

Risk management for quality process should be managed at key points during execution of all quality process activities related to assessment, education, training, selections, and all evaluations. Risk factors can be identified for content related to delivery, goals, and schedules for development. Definition of risk factors is used to solve issues and problems that show struggles in performance at all levels of success. Tool orientation helps to solve many problems to reach expectations that are provided by management.

6.6 Summary

Risk management can be managed and will drive all team members of a hands-on approach and the need for team members' accountability and effective reporting on all status information. Quality Assurance representatives, management, and employees need to provide accurate reports to be evaluated. The information of risk management that is solved will reduce stress and will have common sense for quality control.

Further Reading

Morris, R. A. 2008. "*Everything Project Management*", ISBN 10: 1-59869-635-1.

Pressman, R. S. 2010. "*Software Engineering, A Practitioner's Approach*", New York, McGraw-Hill. ISBN 978-0-07-337597-7.

7
Quality Assurance Improvement with Metrics

Quality Assurance Improvement can be characterized with metrics as a focus of current outcomes and needs to ensure compliance and proper follow-up of identified issues and concerns. While the scope of Quality Assurance improvement with metrics includes actions as conducting a root cause analysis, developing an action plan requires any specific or formal improvement process to be used. When it comes to Quality Assurance improvement, using metrics can make things better and focus on compliance, while performance improvement related to systems issues can cause poor outcomes. Metrics is a standard unit of measuring, assessing, controlling, and selecting processes along with plans and procedures to carry out measurements of all reviewed assessments.

7.1 Quality Assurance Metrics

Quality Assurance metrics are a key component of effective quality management plans and procedures to show that measurements ensure that customers always receive acceptable products or deliverables. Quality metrics are used to directly translate customer needs into acceptable performance measures in both products and processes.

Defining Quality Assurance metrics before starting projects will help you translate your clients' needs into measurable goals. It's critical that you define a set of quality metrics during your project's planning phase so that management and team members know exactly what you need to get done.

Managers must be able to assess the progress, efficiency, improvement, and performance of their projects, and metrics are the means that allow project managers to do this. It is important to note that metrics must be established in an effort to directly improve the processes involved and must be attributable to established goals and customer requirements.

Once all metric measurements are completed, the manager, Quality Assurance representative, and team members will meet to review and compile data and develop their recommendations based on their findings. If any of the metrics have not been satisfied, the manager will include recommendations for correcting the metrics in the Quality Assurance metrics review. This may be a small change to a process parameter or product quality improvement initiatives.

Quality Assurance metric reviews must be scheduled on a monthly basis throughout the program and project life cycle. Management is responsible for scheduling the meetings and ensuring that rooms are reserved as well as making all necessary discussion and visual support. Management is also responsible for ensuring that all required attendees are notified of the meeting in advance.

Quality Assurance metrics are important for all decision points that lead you to take effective actions for improving quality levels depending on metrics that are chosen. Quality Assurance metrics also help judge productivity and efficiency over time, and many test cases that pass in the first run and many that require a retest can lead to numerous reports.

Many metric tools help keep track of metrics and reports on various levels given to insights about project status, milestones, or even individual test runs. All test results for program and project activity are archived so that you can learn from the past, discover trends, and be better prepared with each new release cycle.

7.2 Measure Program and Project Performance

Every manager is given the requirement to provide detailed program and project performance reporting and spending most of their time entering hours worked into work packages on a daily basis. The requirement for that level of reporting is perceived for the manager to find day-to-day activities always and try to measure the program and project's performance.

It is the responsibility of managers to ensure that all programs and projects stay on schedule and within approved budgets. Performance

Table 7.1 Measure Program and Project Performance

	ORGANIZATION CAPABILITIES	POLICIES AND STANDARDS	PROGRAM AND PROJECT PROCESSES	CAPABILITIES
Plan	Mission and vision	Goals, plans, policies, and standards	Processes involved in product life cycle	Access controls for design and internal activities
Document	Roles and responsibilities	Audits and risk management	Performance metrics	System requirements, operating procedures, and performance metrics
Execute	Communication and training	Customer feedback	Improvement and performance	Documented plans, procedures, and policies
Monitor	Review all activities and objectives	Plans and procedures	Process compliance and verification tasks	Reporting on improvement and performance

measurement provides every manager with the visibility to make sure that operations are done within the approved time and cost constraints and that the performance is according to plans and procedures. It also alerts management if program and project activities run behind schedules, so that actions can quickly be taken to get back on track. It is also important to measure program and project activities using Table 7.1.

Developed metrics is based on relationship of important information to relevant companies, institutions, military, and successful businesses to perform the following:

- Process and policy standards
- Management and team member productivity
- Business and company productivity
- Service-level compliance
- Behavior and risk management

The processes, policies, standards, and technologies are required to manage and ensure the availability, accessibility, quality, consistency, auditability, and security of data within the organization by using effective metric evaluations. The accuracy, completeness, and validity of the data meet or exceed user expectations by using metrics for measurement of program and project performance.

With management sponsorship, pilot the identification of business objectives using metrics that are impacted by the quality of data. For each of those objectives:

- Determine improvement and performance metrics to be communicated.
- Have expectations for achieving objectives.
- Review rules with management for agreement and sign off on key metrics.
- Apply rules using proper tools to assess baseline metrics.

Using proper metric tools and methods will support Quality Assurance monitoring, audits, and improvement within processes and for ongoing improvement and performance objectives.

7.3 Using Metrics for Quality Assurance

While working with Boeing Defense and Space for 15 years, I was performing audits, reviews, and evaluations daily and was working with a metrics database for all problems, concerns, and issues discovered during my software quality activities. The metrics database was created by Earick Gamble, who was a Boeing co-worker, and we worked together using metrics for reviews, audits and evaluations. It was important and required tasks to use metrics to work with Boeing management, employees, and government representatives. The metrics that I documented were fixed to ensure that all audits, reviews, and evaluations are complete and closed for compliance.

Identifying the metrics establishes the operational direction by enabling programs to prioritize the information policies in relation to the risk for metrics impact by listing the expectations for acceptable audits, reviews, and evaluation performed for measurements that allow acceptability thresholds for those emerging metrics. By listing the critical expectations per metrics, methods for measurement and specifying thresholds can associate data with levels of success in activities performed.

The success criteria should be noted in relation to the ways that Quality Assurance improvement time was spent on corrections and to show success by increasing the speed of delivering information as well as increasing major decisions and achievements or milestones for success criteria that allows management to provide individual accountability and reward achievement.

Quality Assurance metrics are for processes to be in place for operational performance, depending on measuring conformance to expectations and knowing that appropriate data requires two things:

- Method for quantifying conformance.
- Threshold for acceptability.

Policies drive the way organization conformance is related to information policies and the way that information policies impose rules that can be monitored throughout processing activities. Improvement and performance objectives center on maximizing productivity and reducing organizational risks that are needed to impose or manage the way that programs and projects are performed, data definitions, information policies, and data structures and formats.

7.4 Software Metrics

Software metrics is a measure of software development and implementation that is qualified and compliant. Software metrics are important for many reasons, including measuring software performance, planning work items, and measuring productivity. In the software development processes, there are many metrics that are related to each other. Software metrics are related to planning, organization, control, and improvement. Topics include many examples of software metrics:

- Benefits of software metrics
- How software metrics lack clarity
- How to track software metrics
- Examples of software metrics

Metrics is fundamental to software engineering as a discipline. With software measurement, a system is assessed using a range of metrics, and from these measurements, a value of the system quality can be implemented. Software metrics can be either control metrics or predictor metrics. Metrics may be used to control the software process or to predict product attributes. Metrics can control the following factors that affect software product quality:

- Process quality for activities related to the production of software, tasks, or milestones.
- Product quality results of the software development activity, deliverables, and products.

Product metrics help software engineers to better understand the attributes of models and assess the quality of the software, and they help software engineers to gain insight into the design and construction of the software. Software engineers focus on specific attributes of software engineering work products resulting from analysis, design, coding, and testing and provide a systematic way to assess quality based on a set of clearly defined rules and insight into software development.

Measuring the length of code using number of lines of code or assess software quality through defects/bugs. A line of code is any line of program text that is not a comment or a blank line, regardless of the number of statements or fragments of statements on the line. This specifically includes all lines containing program headers, declarations, and executable and nonexecutable statements. Accepted measurements are dependent on the programming language and programmer and well designed with short programs. Source code creation is only a small part of the total development effort, and it is often unclear how to count lines of code implemented for delivery to the client when the product is completely finished. Most work on software measurement has focused on code-based metrics and plan-driven development processes, and more software is now developed by configuring system requirements. Software measurement per metrics can be used to gather data about software and software processes. Product Quality Assurance metrics are particularly useful for highlighting anomalous components that may have quality problems.

7.5 Metrics Used for Success

Companies, institutions, military, and successful businesses have goals to make sure that metrics are used for success and provide many quantified questions to implement metrics. Metrics should be used to answer questions that support specific goals that should be done as long as the questions and answers help drive positive changes.

All programs and projects have some set of invariant goals, questions, and accomplishments using metrics for success. Measurement of activities, such as user engagement, development, and so on, provides feedback on how the programs and projects are doing in the real world. Changes to development and implementation will also affect these kinds of metrics. There are many objective metrics that should be monitored continuously to make incremental improvements to processes and production environments. Improvements in these metrics will guarantee that your customer satisfaction levels will rise by leaps and bounds.

For Agile and Lean processes, the basic metrics are lead time, cycle time, and team velocity. These metrics aid in planning and inform decisions about process improvement, and they do measure success and provide added value to do the objective Quality Assurance you should measure. I'll explain why below:

- **Lead Time**: If you want to be more responsive, work to reduce your lead time by simplifying decision making and reducing wait time, which includes cycle time.
- **Cycle Time**: Using continuous delivery can have cycle times measured in minutes or even seconds instead of months.
- **Team Velocity**: Metrics should only be used to plan iterations because metrics is based on nonobjective estimates. Treating velocity as a success measure is appropriate and making a specific velocity into a goal provides value for estimation and planning.

When metrics are out of the expected range or trending in alarming directions, talk to the teams and get the whole story, and let the team decide whether there is cause for concern and how to fix problems. Metrics give insight into where important and essential processes need attention. A high open rate and a low close rate across a few iterations mean that the production issues are currently a priority that the team is focused on reducing technical issues and how to fix them quit.

The joy of using automated tools for tracking and measuring quality metrics frees up time to focus on the metrics that really matter for success.

Quality-based metrics relates to performance to provide the good service to the specified levels that can be measured in terms through technically defined responsibilities and activities. There are standard metrics that should be incorporated as components that can be built into a service-level agreement. Programs and projects can be assessed through a range of measurement and metric tools to ensure that system development by selected metrics reflects the requirements to do its best to meet all required tasks.

7.6 Corrective Actions with Metrics

For corrective actions with metrics, I recommend following eight disciplined step approaches to resolving internal and external quality issues. The steps for problem-solving are as follows:

- Management and team members approach.
- Describe and address problem.
- Containment action for ensuring that problems will be fixed.
- Root cause verification and validation.
- Implement corrective action using metrics.
- Verify corrective actions to fix problems.
- Prevent recurrence and future issues and problems.
- Management and Team members are given thanks for fixing problems.

Corrective actions with metrics normally take some time using problem-solving method steps. Complete the corrective actions in a reasonable amount of time to satisfy your customer and keep the customer from walking away. Management, team members, and Quality Assurance representatives must verify the corrective actions by measuring with metrics and monitoring the results after implementing the corrective actions. Make sure problems cannot be recreated, and Verification and validation include reviewing documentation that supports the process changes from the corrective action. Complete the verification and validation to implement the corrective action. Identify the root cause and always take the appropriate corrective action to fix it. It is important to list all possible corrective actions and using metrics, as these will depend on all situations.

Preventive actions pursue steps to prevent issues and concerns from recurring in the future. Possible preventive actions include examination of all production lines and implement corrective action as necessary to schedule periodic training for the corrective actions. You must schedule periodic audits for the problem and corrective action activities and include additional reviews for the problems and update all quality plans.

Companies, institutions, military, and successful businesses should have a documented system that describes the methods employees need to follow during the corrective action problem resolution using metrics.

It takes significant effort to resolve many problems. Management will always need to congratulate the team members and Quality Assurance representatives. This encourages involvement on future problems and don't document the congratulated actions or share these with your customer.

Throughout my life working as a software engineer and focused on Quality Assurance, I went through audits, reviews, evaluations, verification, and validation using metrics to document problems that need to be fixed. Once I started my own business as BL Summers Consulting LLC, I became effective. I always have a guide for setting or achieving my goals, so that I can bring results.

Much of the consulting work I do is talking to business people who fail to bring in results in their businesses. I ask them about their goal-setting processes, and almost every person responds with I don't set goals, because I never hit them. The major breakthroughs for companies, institutions, military, and successful businesses happen when they understand clear decisions and are focused on actions for the result of simply having clear goals.

It is important to expect success and describe what success means. This is a huge step in making what you think, see, and feel for people who write down their thoughts and gain better insight in thinking processes for readily accessing information when needed. Once there is clear vision and a know-how to set goals, companies, institutions, military, and successful businesses, simple steps can be taken to achieving your vision.

Further Reading

Broughton, R. 2018. Quality Assurance Solutions. Online information.

Loshin, D. 2006. *"Business Performance and Data Quality Metrics"*, Silver Spring, MD, Knowledge Integrity, Inc., https://www.quality-assurance-solutions.com/

Lowe, S. A. 2018. Software Developer and Principal Developer and Consultant at Thought Works, Head-First Domain – Driven Design, O'Reilly. https://saturn2017.sched.com/speaker/steven_a_lowe.1wbea8ql

8
Quality Assurance Framework

The Quality Assurance Framework (QAF) defines important practices for the following:

- Project management
- Process management
- Companies and institutions
- Engineering
- Support

Each practice comprises multiple areas of Quality Assurance practices. In the QAF, a set of important practices are defined for areas that are relevant to acquisition and artifacts documented for processes used for Quality Assurance evaluations and must be documented for each important practice.

8.1 Required Evidence

The Quality Assurance auditor must always look for evidence to what practice is implemented by examining artifacts that are related. By improving the QAF, go beyond identifying defects and issues towards the end of the life cycle and evaluate how QAF decisions can be made during evaluations and audits for requirements affected by known defects and issues found for removal of these issues and concerns.

A formal review and examining requires identifying potential associated with operational risks identified. Quality Assurance auditors must always document the rationale behind QAF and provide evidence that support the rationale. Metrics that is documented in Chapter 7 provides evidence that justifies the Quality Assurance associated with important required decisions for justified measurements.

Table 8.1 QAF Questions

Executions	How good were all product and process management activities?
Effectiveness	Were effective Quality Assurance reviews, audits and evaluations incorporated in all required practices?
Results	Were all results that were analyzed incorporated and discussed?

Quality Assurance case is a documented body of evidence that provides a valid description that can be specified as critical claims to ensure QAF practices are justified in an improving environment and having effective measurement for required planning to reveal what measurements reveal. Tracking all results can aid in understanding and justifies effective measurements required for planning of efforts achieved for all intended outcomes. Metrics provide help to QAF and provide answers to questions shown in Table 8.1.

8.2 Requirements and Risk Mitigations

In some instances, QAF risk mitigation is dynamically applied to issues and concerns with management and team members to ensure that insecure practices. Outputs include planning, working tasks, and results. Quality Assurance auditors can show positive terms in what requirements performed for authentication, authorization, and having positive requirements implemented verified that hold true and confirming results that meets plans, processes, engineering, and company contract support. Negative requirements that cause risk mitigations should never occur. Requirements and risk mitigations target the weakness and mitigations identified by analysis and will confirm the level of confidence companies should have and must be fixed.

Requirements planning can become a workload for companies, institutions, military, and successful businesses based on planned processes and documented procedures. Planning can become a long-range activity, providing requirements to solve critical issues and concerns. The planning for requirements should provide daily reporting and can utilize specific lists of critical resources for project and product development and the ability to define standards for efficient delivery to customers knowing that the processes were successful.

8.3 Using the QAF for Companies and Institutions

Using the QAF provides best practices and can create Quality Assurance goals to provide evidence of how goals are addressed through QAF practices. At each Quality Assurance audits always show activity progress, outputs, and metrics to be reviewed and evaluated to confirm that progress is addressed and evaluated to reach compliance for all activities. Each audit will ensure that companies and institutions follow results that are supported by the following evidence:

- Assess capability needs.
- Ensure solutions will be cost-effective and operational effective.
- Make sure all system requirements are defined and consistent.

8.4 Benchmarking Process

Benchmarking Process requires Quality Assurance representatives to provide oversight to performance conducted for all companies, institutions, military, and successful businesses and to identify and act upon all processes requiring improvement. All Benchmarking Processes involve a systematic collection of data with a view to making relevant comparisons of aspects of processes, performance, and outcomes for success with oversight for the purpose of improvement as stated later:

- Identify strength and weakness in all performance.
- Obtain all provided data for supporting decisions.
- Make sure to determine actions to improve processes and increase performance.
- Measure and compare to make sure processes are doing better.

The relevant activities Quality Assurance provides for oversight is teaching and development to determine the benchmarking processes to specify subjects, topics, activities, and outcome for benchmarking. Benchmarking processes will include required data and evaluate the results collected. The final report is presented to the management, which includes the recommendations for improvement and changes for consideration and endorsement.

Figure 8.1 Quality Assurance Framework.

8.5 Benefits of QAFs

Benefits of QAFs can gain more effective and more efficient plans and procedures to provide better quality services for team members and employees to provide better communications with management. There should be more creative thinking, enabling new perspectives and ways of working, and provide continuous improvement over time. Refer to Figure 8.1.

I would like to bring up Six Sigma as a set of techniques and tools for process improvement. Six Sigma seeks to improve the quality of processes by identifying and removing the causes of defects and also uses a set of Quality Assurance methods and creates a special infrastructure within the organization who are experts in these methods. Each Six Sigma project carried out follows a defined sequence of steps and has specific value targets, for example:

- reduces process cycle time
- reduces pollution
- reduces costs
- increases customer satisfaction
- increases profits

Six Sigma asserts continuous efforts to achieve stable and predictable process results for the importance to business success and processes that can be defined, measured, analyzed, improved, and controlled. Achieving sustained quality improvement requires commitment, particularly from top-level management. Six Sigma implements goals to improve all processes that need to determine an appropriate sigma level for each of their most important processes and strive to achieve all process development. As a result of this goal, it is incumbent on management to prioritize areas of improvement.

8.6 Summary

The QAF will be successful when everything is looked at and evaluated to meet compliance for all Management and Team employee activities. The guidelines must always be used to enforce project management, process management, companies and institutions, and support implemented and working to ensure all tasks and activities are meeting expectations.

Further Reading

Alberts, C. J. 2017. "*Prototype Software Assurance Framework (SAF): Introduction and Overview*", Pittsburgh, PA, Carnegie Mellon University.

Toro, C. 2011. Master Black Belt, PMP, MBA Lean Six Sigma Toolbox, Consultant for Businesses needing effective business roles and Quality Assurance disciplines.

9

Quality Assurance Process Improvement

There are many types of Quality Assurance Process Improvement roles for supporting the process that will be documented in this chapter. Companies and institutions must have management infrastructure first and second is the team employees. The first covers all roles and responsibilities, and the second covers operational tools, facilities, and customers. All these roles need to be active in supporting process improvement for Quality Assurance.

9.1 Process Improvement Direction

There should always be a model for defining Quality Assurance process improvement with a step-by-step direction toward effective improvement. The process improvement stages and levels can be achieved by reaching those levels and using an example such as the Capability Maturity Integration Model or other levels to match companies and institution needs. The results of making assessments should always identify the strengths and weaknesses with Quality Assurance personnel for improving the processes to be effective and compliant. There are symptoms of having effective process improvement such as

- Discipline will ensure that management and team members follow processes as a norm.
- Enforcement is effective by making sure that conformance is checked and issues reported.
- Visibility of process documentation can be effective in supporting management.
- Measurement and evaluations of performance are linked to process goals.
- Training is mandatory for the team members when new members join the team.

9.2 Quality Assurance Process Infrastructure

Quality Process Infrastructure is a very important activity and a basic structural foundation as an underlying framework for companies and institutions including policies, standards, training, and applicable tools needed for supporting day-to-day performance. In order to establish an effective infrastructure, the Quality Assurance process environment must have in place a management, organizational, and technical infrastructure. The roles and responsibilities must be operated at a corporate and project level to ensure all roles and responsibilities are covered in a management structure. The management role provides guidance and strategies for Quality Assurance Process Improvement activities such as monitoring progress and fixing issues and concerns.

I would suggest as a Quality Assurance consultant that companies and institution put together a process group for providing coordination and guidance to team members in order to achieve all goals and high expectations. The process group would be established under management and ensures that the right people are selected to help fix and resolve any problems addressed.

Infrastructure is needed to enable and facilitate Quality Assurance processes and to support all roles and responsibilities and covers the following aspects:

- Roles and responsibilities have to be in place to sponsor, manage, and perform.
- Necessary technology tools are necessary for facilities to support process improvement.
- Monitoring all process improvement activities.

9.3 Effective Process Environment

An effective process environment reflects symptoms and behaviors that are visible all processes, roles and responsibilities, management support for following plans and procedures. Training and continuous process improvement are important to ensure critical factors treat improvement plans as a full-scale project and have commitment to all resources, project management, plans, schedules, and continuous improvement.

Management roles and responsibilities support effective operations for team member roles with respect to responsibilities documented related to templates and audit checklists that are applicable. The world is full of process and product plans and what can be thought of as an effective process environment. Process performance goals must always be aligned with companies and institutions to bring compliance and increase effective process activities. Audits of process results ensure that process benefits are handled to provide enforcement against any nonconformance activities found and provide to management the commitment that processes are visible and are required to achieve process culture. In the absence of process culture, the process environment will be more isolated to all roles and responsibilities. An effective process environment, it should have the following roles and responsibilities:

- Process training and ownership.
- Measurement and metrics of process results.
- Reviews and audits of process performance.
- Provide feedback from all team employees.
- Process inspection and enforcement.

9.4 Document Action Plan

Companies, institutions, military programs, and successful business companies should take an approach to converting Quality Assurance processes into an improvement action plan to specify steps to be followed to develop process improvement. All findings should highlight the strengths and weaknesses of processes that are being looked at and analyzed. Assessment findings could be converted into process improvement, as stated in Figure 9.1.

9.5 Quality Assurance Process Execution

Quality Assurance Process Execution is where tasks, activities, and work will get done. Managers must communicate with team members and employees on a regular basis to ensure that results are provided on a daily basis. Receiving the right answers during process execution is required:

- Will the tasks and activities be done per the date provided?
- What is the percentage of all tasks and activities that are completed?

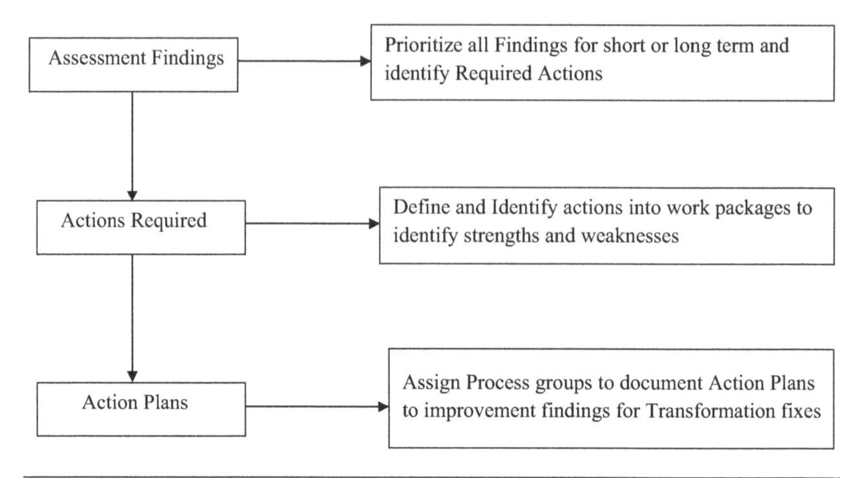

Figure 9.1 Assessment findings.

Companies, institutions, military ,and successful businesses will get answers if automated questions are asked to ensure processes used are at a correct pace to identify potential issues that will not derail success. Quality Assurance representatives (QARs) should gather information on the progress from management, team members, and employees to form a bond by working together.

If everything is going according to plan, thank everyone for their efforts and the overall success being worked. Plans must be set appropriately, and communication will always be the key in letting everyone know what is coming to help move process execution to be working as a success. Process control goes hand in hand with all planning and execution phases, where direction will react to information, understanding processes and monitoring to make sure all tasks and activities are progressing as planned. There are three adjustments that can be made in planning to shorten tasks and activities:

- **Fast Track**: Increase schedules and work parallel to shorten time frame.
- **Crash**: Schedules can add resources to shorten time frame.
- **Remove Requirements**: Tasks and activities from project development.

The product life-cycle discussion could occur in conversations provided by management, team members, QARs, employees, and suppliers, but management must move throughout all phases such as

- Request the status of start and completion for executions.
- Control and assess the impact of status and make proper adjustments.
- Planning new resources adjusting time, duration and cost impact.

9.6 Quality Assurance Representatives

QARs must be bright and shining as a superstar, such as teams who win major sport championships. The bottom line is that well-organized QARs will get the job done right! The right QAR must be reliable, trustworthy, flexible, get along with everyone, make sure rules are understood, and be there when help is needed. Managers seeking QARs should make sure that qualifications may be necessary so that tasks and activities are performed per process evaluation and have a fill for understanding all Quality Assurance responsibilities. The following items are important when selecting QARs:

- General experience is important and valuable.
- High profile and accomplishments.
- Worked and helped other companies, institutions, military, and successful businesses.
- Technology skills are great to have.
- Trustworthy for dedication and commitment.
- Make sure respect is used during support.
- Clear lines of communication.

Further Reading

Morris, R. A. 2008. "*The Everything Project Management Book*", 2nd Edition. New York, Simon & Schuster. ISBN10:1-59869-635-1.

Zahran, S. 1998. "*Software Process Improvement: Practical Guidelines for Business Success*", Boston, MA, Addison-Wesley. ISBN 0-201-17782-X.

10 Effective Processes for Quality Assurance

As companies, institutions, military programs, and successful businesses move forward, it is important and mandatory to monitor the effective process of Quality Assurance based on the number of topics, including time frame, cost, and performance. Monitoring could show the highlight of potential issues and problems discovered, and there will be a communication to management and team members on exactly where programs and projects stand and how closely monitoring will resemble reality. Consistent monitoring helps avoid a number of disasters and could protect all concerns before reaching the required goals and schedules.

10.1 Success of Programs and Projects

There are many questions that could be asked per monitoring and knowing progress is becoming successful and changes are implemented when needed to become complete and compliant. As programs and projects become a success, keep these questions discussed:

- Are we reaching and meeting accomplishments for working schedules?
- Are we working toward all goals for accomplishments?
- Are there warning signs for impending issues and problems?
- Is there pressure from management to complete program and projects early?
- Is there opposition to the program and projects to be completed?

To keep progress plans and data on a daily or weekly basis, always have frequent meetings to show how often monitoring reaches the scope of the program and projects to see the number of people working to improve processes. Bigger and larger programs and projects will require more monitoring, since some areas will fall or more problems could arise. Quality Assurance auditors need to know all people working at all companies and institutions and also outside contractors, suppliers and others to help complete monitoring tasks and activities. When more people are involved, staying on track will be better for monitoring of processes for Quality Assurance and take the advantage in aspects of programs and projects. Make it clear to the people that they can provide questions and problems they see when being trained or are new to defined tasks.

10.2 Communication Needs

If managers and team members expect daily, weekly, and monthly updates along with audit reports, communication needs are important to ensure that latest developments are addressed. When there are high-profile programs and projects, it will require managers and team members to provide updates to many questions being asked. When Quality Assurance information is gathered, the information must be addressed to determine what is to be gathered:

- When was each program and project started and completed?
- What resources are used for each activity?
- The number of hours put into for each activity.
- The goals of each activity being completed or accomplished.

10.3 CMMI Guidelines for Process and Product Improvement

The Capability Maturity Model Integration (CMMI) collects the best practices the help companies and institutions to improve process and product improvement. The best for CMMI will always focus on all activities for the development of quality products and service for meeting the needs for customers and all end users. The development of CMMI requires a team or group to provide consultation on many CMMI project issues and concerns to help improve all activities for improvement to process and product improvements.

All companies, institutions, military programs, and successful businesses companies will always want to deliver and support services better, cost reductions, process and compliance increasing complex process and products. Management and team members must be able to manage and control complex and maintenance activities. CMMI provides an opportunity to eliminate numerous issues and problems and will provide best practices that will address all process and product services.

Models and standards is an international standard that specifies the general requirements for the compliance to carry out tests and or deliveries. These requirements outline what a laboratory must do to become accredited and successful. Management systems refers to the organization's structure for managing its processes or activities that transform inputs of resources into a product or service which meets the organization's objectives, such as satisfying the customer's quality requirements, complying with regulations, or meeting environmental objectives. The CMMI model is widely used to implement process and product Quality Assurance in companies and organizations. The CMMI maturity levels can be divided into important steps, which achieve by performing specific activities within companies and organizations.

10.4 Process Improvement Services

What holds process improvement services together is the research to help companies, institutions, military programs, and successful businesses and the focus on all employees, plans, procedures, methods to ensure that all developed tools and equipment are working. What holds everything together is the process used for business, which is shown in Figure 10.1.

10.5 Decision Analysis and Compliance

The purpose of Decision Analysis and Compliance is to review decisions used on audits and evaluations to monitor processes to identify established concepts and expectations. Guidelines are used to

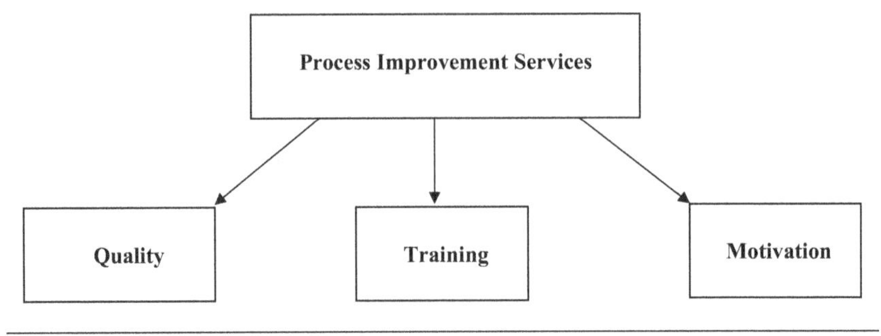

Figure 10.1 Process used for business.

determine which issues and concerns are noncompliant and meet the requirements for process and performance. A formal Quality Assurance audit and evaluation involves the following actions:

- Identify and fix alternative solutions.
- Establish the process and plan and policies for evaluations.
- Provide methods for the evaluation plan.
- Select solutions to provide higher process improvement and show compliance.

Plans and policies are helpful when performing Quality Assurance audits and reviews and to define company and institution expectations and is visible to management and team members. Senior management is responsible for guiding principles, communications, directions, and high expectations. Policies establish addressing to meet outcomes and maintaining and controlling products that are changed to make sure integrity is implemented. There are always expectations that a policy will implement for Decision Analysis and Compliance and will provide guidance for all decisions that are performed by management and team members.

Process and product management are provided by the managers and team members with capability to document plans and provide the best practices, learning, and training. All activities include process and product improvement for measurement and lessons learned to make sure these topics are implementing and working for compliance to requirements.

The basic process and product areas are shown in Figure 10.2 for management.

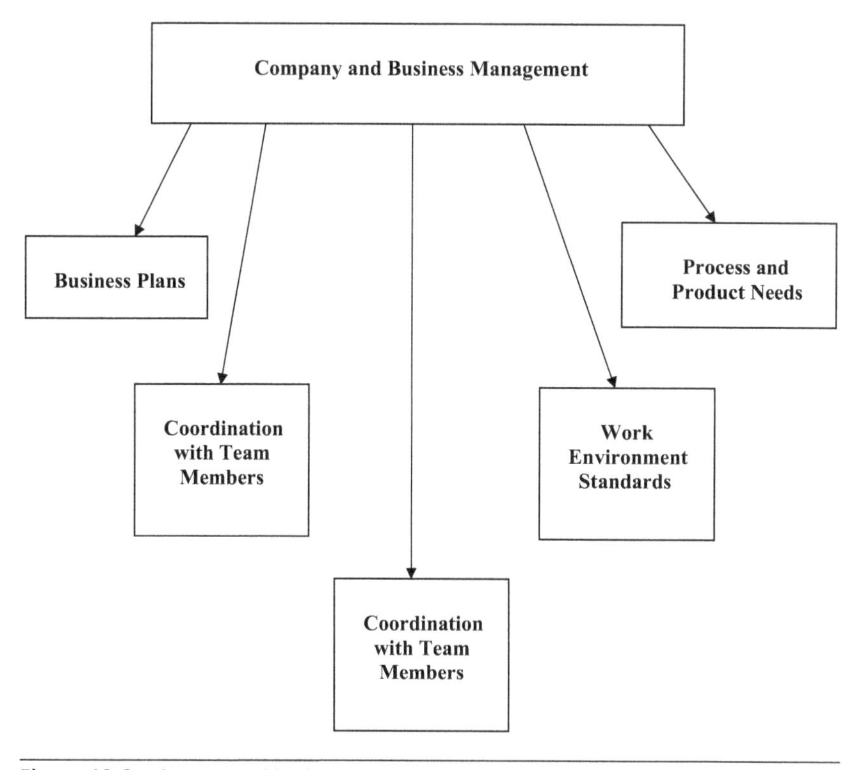

Figure 10.2 Company and business management.

Further Reading

Chrissis, M. B. 2011. "*CMMI for Development: Guidelines for Process Integration and Product Improvement*", 3rd Edition. Boston, MA, Addison-Wesley. ISBN 978-0-321-71150-2.

Conrad, M. 2011. "*CMMI for Development: Guidelines for Process Integration and Product Improvement*", 3rd Edition. QA76.758. C518. 005. 1-dc22.

11 QUANTITATIVE PROCESS PERFORMANCE AND COMMITMENTS

The purpose for Quantitative Process Commitments is to control processes performed by companies and institutions that represents results that are accomplished and compliant for goals of performance that are defined and described in key areas that can provide analysis, measurements, and acceptable process activation. When process performance is accepted, plans, projects, and tasks are established, and they can be used to control process performance and commitments.

11.1 Goals and Process Capability

The Goals and Process Capability describes the expected results from the following key process areas (i.e., achievable outcomes that are expected from current and further projects to be accomplished to be successful). These Goals and Process Capabilities must be used to establish process performance and to analyze all defined process and expectations. When you look at the goals that have expectation, you see the following:

- All quantitative processes are planned.
- Processes for performance are defined and controlled quantitatively.
- The process capability of companies and institutions standards is known in quantitative terms.

11.2 Perform Commitments

All projects follow documented policies for measuring and quantitatively control defined by the processes documented in plans and procedures. When you look at quantitative control, the term implies is based on techniques that can analyze processes, identify variation in performance, and can bring performance in well-defined expectations. Companies, institutions, military programs, and successful businesses must follow written policies to specify measurements and maintain process capabilities tied to stand processes. The process capabilities include the following:

- Description of plans and procedures that are standard.
- Standard definitions of measurements.
- Expected values for accomplishments related to plans, procedures, and measurements.
- Support required to record and analyzing data selected for processes and commitments.

Maintaining commitments is important when processes are developed and documented in plans and procedures that specify strategic goals for project goals to qualify and show productivity during life cycle time. Management and control implies that work products are given time and all changes are used and implemented in a control manner.

Quality Assurance representatives implement and support process management required for training to model and analyze all process performance and commitments by selecting, collecting, verifying, and validating data used for measurements. Plans and procedures specify the companies and institutions' product quality, productivity, and standard process goals for quality, productivity, and product compliance and also make sure plans are managed.

The applicable Quality Assurance commitments provide an overview of the capability in terms of relevant operations, and the time frame under consideration describes the commitments and capabilities that all programs and projects deliver and how it relates to applicable commitments to required capability. Assets required to achieve commitments describe the types and initial quantities of assets required to attain Quality Assurance processes and to identify and define the initial asset quantities needed to achieve commitments.

11.3 Verification and Validation

The activities for verification and validation performed by Quality Assurance representatives are then reviewed by management and team members on a periodic basis. The audits and reviews performed cover the content of product management oversight and in preparation of verification and validation tasks and activities will apply to the following:

- Required data is collected, needed, and does exist.
- Collected data provides support for goals and objectives.
- All data is accurate, correct, and compliant.
- The data could be confidential and properly protected.

Accurate reviews and audits are essential to the companies and institutions to define the framework and specific requirements for **verification** and **validation** of product development and efforts. The Quality Assurance that products delivered to customers have been reviewed, audited, verified, and validated meet the required quality requirements. The verification and validation begin during concept definition and will always continue through requirements of deliverable products. Verification and validation address work products in fine environments against all secreted requirements for customers and product element requirements. Companies and Institutions must develop top-level verification and validation plans that will establish common practices that are set for verification and validation to analyze and ensure results are accurate and compliant. The following tasks must be performed as follows:

- Analyze data from verification and validation to show expected results.
- Ensure correct standards of products submitted for verification and validation.
- Document results for each activity.
- Coordinate verification and validation results with users and team members.
- Enter change process with any new or modified requirements that are reviewed.

Product reviews and controlled requirements indicate that verification and validation are to be performed to ensure acceptability by customers and, if not accepted, must be fixed and approved by team members involved to ensure that all products are ready for use and integrated to meet all requirements and compliance before delivery to customers. After verification and validation reports are developed and written and the requirements are determined, prepare compliance for closure to plans and documentation to show proof of compliance.

This book will describe the effective processes for Quality Assurance activities performed in support of verification and validation. The purpose of verification is to ensure that the product meets its specified requirements and is accomplished through various activities ranging from in-process reviews and audits. The purpose of validation is to ensure that the work product fulfils its intended use when placed in its intended environment. Quality Assurance representatives will audit all activities and support qualification by monitoring all activities.

11.4 Quality Engineering Knowledge

Quality Engineering Knowledge comprises structured processes ranging from requirements specifications, standards, test reports, enterprise architecture models to system configurations. Models play an important role in mapping this knowledge. Quality Engineering Knowledge is also based on the generated Quality Assurance processes that is made available with tool-based technology.

Prime importance is the focus on *Quality Assurance* tasks for recognition of risks, and appropriate support for companies, institutions, military programs, and successful businesses and to show results in the following requirements for a Quality Engineering Knowledge based on

- Important quality criteria include knowledge that is consistent and up-to-date as well as complete and adequate in terms of relation to tasks and activities.
- In context, organizations interact with the Quality Engineering Knowledge base to provide mechanisms for ensuring confidentiality and integrity.
- Quality Engineering Knowledge bases offer a whole range of possibilities for analysis and finding information to support quality control tasks.

11.5 Quality Assurance Process Performance

The Quality Assurance Process Performance provides a production context for Process Performance Qualification (PPQ) as an element of Process Qualification. The process for Qualification is the PPQ that is part of the process life cycle that includes protocol, execution, and final data reports. The standards expect to verify that the tools, equipment, facilities, and utilities are qualified for use. Furthermore, the final data reports of the PPQ stages demand that recommendations of the Quality Assurance tasks and activities are required for current and future development.

The data reports are collected and evaluated for performance and acceptance criteria for each significant processing stage per plans and procedures. Criteria and process performance and the qualification of tools and equipment, training qualification, verification and validation sources are important for analytical methods for approval and buy-offs by Quality Assurance representatives.

Quality Assurance Process Performance is important in any organization, since it determines the future growth of an employee. Performance reviews help in guiding people to be responsible for documenting performance evaluations to effectively appraise and report all assessments.

Data reports ensure that feedback can be given to the management so as to encourage them to perform better. Organizations with team members and employees are responsible for documenting and conducting performance reviews along with the reporting results.

Companies, institutions, military programs, and successful businesses will ask the team members and employees to conduct the appraisals solely based on management judgment. Performance evaluations help for phrasing performance by giving insight into evaluation comments.

It is important to highlight the positive ways in which the team members and employees contribute for development, proper communication, improving motivation, organizational targets, and ensuring that positive relations are maintained with management. Effective performance review tips need to be kept in mind while documenting data reports for use that can support in the process performance tasks and activities stated as follows:

- Communications skills meet the requirements.
- Managing people meets the requirements.
- Leadership needs improvement.

- Teamwork exceeds the requirements for building a strong team.
- Delegation meets the requirements based on their skills, experience, strengths, and limitations.

Further Reading

Humphrey, W. S. 1995 and 1989. "*A Discipline for Software Engineering and Managing the Software Process*", Boston, MA: Addison-Wesley.

12
Understanding the Quality Assurance Direction

Companies, institutions, military programs, and successful business strategies and defined corporate policies establish the requirements and boundaries for correct and compliant information and directions that provide management, departments, focus, value, and leadership. All strategies could have many different approaches, and it is important to utilize information that can provide companies to be competitive and take advantage for understanding the Quality Assurance direction.

If you don't know where you are going, you might wind up someplace else.

Yogi Berra (1925)
New York Yankees baseball catcher

12.1 High-Level Quality Assurance Direction

There are many companies, institutions, military programs, and successful businesses that should be integrated with high-level Quality Assurance direction for planning processes to go fast or quick. Many surveys have been provided to make sure management reviews daily how information systems, research, development, and human resources are understood to meet company's mission, vision, and values. It is important to document high-level Quality Assurance direction to define the direction needed for business direction and to be successful and to show strength and not weakness.

If businesses and institutions do a great job of strategic planning, they can focus on the goals of the business to ensure that everyone understands business direction and the challenges that exist. The following steps later can provide the direction needed:

- Review strategic plans and information regarding effective planning.
- Management meet with team members to ensure that questions are answered.
- Meetings are held to report Quality Assurance directions that are working.
- Assemble all information in a summary report.

Many companies, institutions, military programs, and successful businesses will summarize strategic planning by reviewing and understanding the key factors for success, key issues, and conditions that will affect information on missions and directions. I have always felt that having a high-level Quality Assurance direction will help business plans to be working and compliant to all requirements implemented and ensure that effective operations are improved and processes are followed. There are many goals for high-level Quality Assurance directions that business and institutions can use for current and future missions performed to show strength and eliminate weakness for all business plans.

12.2 Roles and Responsibilities

It is time for team members to have a strong understanding of how important their roles and responsibilities and lead implementation efforts and lead become strong leaders for companies, institutions, military programs, and successful businesses. Communication and planning skills are important during the selection phase of plans and projects that all company employees participate in. It is important for the team members to have a full-time involvement and commitment instead of part-time support for all activities of implementation of program selection and to ensure that all activities are compliant and work well. Team members must understand project planning and show continuous communication that is essential to show interest for all activities being performed.

The focus of program and project missions is to rely on scope, goals, and roles and responsibilities of each team member to participate and provide communication and results to management to show status and any issues. Schedules established by management must be accomplished and working to ensure successful tasks are being performed. A Quality Assurance representative is important when decisions are being made for project communication of processes that are being performed and knowing that schedules are being met.

Team members is a collective term for various types of activities used to enhance social relations and define roles involving collaborative tasks. It is distinct from team training, which is designed by a combination of managers for learning and development and to improve efficiency.

Many tasks and duties aim to expose and address problems within business companies, institutions, and military programs. Over time, these activities are intended to improve performance in a team-based environment for team building as one of the foundations of organizational development to be applied to groups such as employees, customers, and inside and outside organizations. The definition of team member roles will include the following:

- Understanding the goals needed for improvement.
- Building effective working relationships.
- Finding solutions for all team problems.

The Quality Assurance representative can assist on any negotiations when needed with outside suppliers to ensure contracts are being followed and meeting all technical requirements for software and hardware installations that fit well for the success of a business.

The following team member project communication is important as shown in Figure 12.1.

12.3 Effective Methods for Quality Assurance Direction

Effective Methods for Quality Assurance direction provide an understanding and importance of critical factors such as planning, design, requirements, configuration management, integration, testing, subcontractors, and quality. Many companies, institutions, military programs, and successful businesses program and project design, build, and test work products effectively and provide the framework of disciplines during development life cycles. These methods support

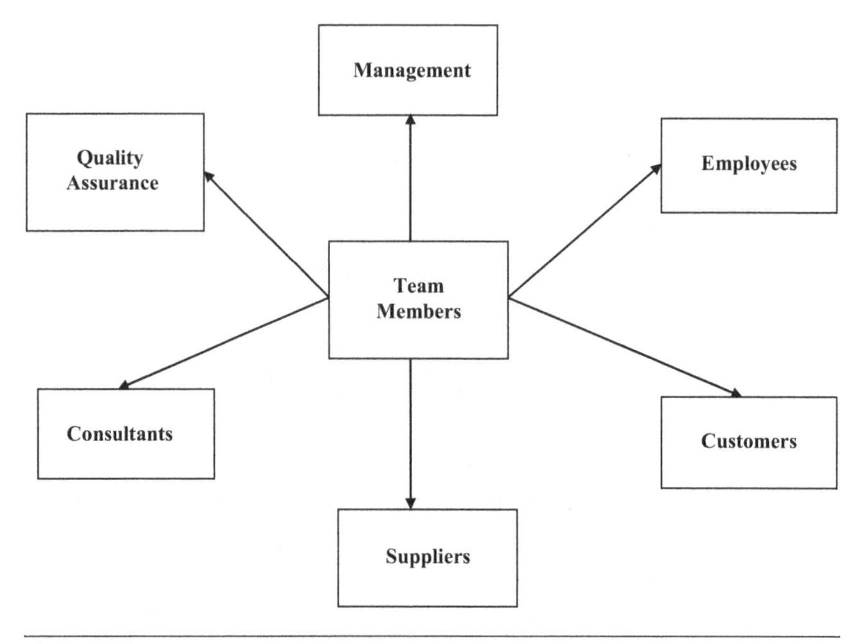

Figure 12.1 Team members.

the building of baselines inside integration environments to prepare for delivery of effective and compliant products to customers.

The primary purpose for the implementation of Effective Methods for Quality Assurance direction increases communication, knowledge, and the visibility into the software life cycle of operations. I hope that you find this chapter informative, interesting, and convey that these methods are more effective for future businesses and institutions. Companies could benefit as well by adopting these effective Quality Assurance methods and directions.

Important Quality Assurance disciplines are defined and discussed throughout this book. These disciplines implemented programs and projects to improve the critical advancement into the phase and practice of being more effective during management-scheduled activities. In order to develop, operate, and maintain Quality Assurance capabilities and directions inside facilities and production, there must be a major discipline in supporting the entire activities (i.e., planning, requirements, design, builds, installations, integration, subcontractors, quality, and delivery) need to be completely understood. The critical understanding and the start of the right disciplines of these methods will empower and achieve effective, flexible, and quality results in all work environments.

The Effective Quality Assurance methods and directions are an integral part scheduled and performed by Quality Assurance personnel for program and/or project-level activities on an on-going basis. These methods and directions form the basis for certification that development activities have been performed in accordance with program and project plans and procedures that are in line with required quality requirements. At times, there are problems with the delivery of work products to customers. The Quality Assurance representative must solve process issues and concerns and participate in reviews and audits and witness all tasks and activities. The Quality Assurance representative must provide assistance and help program and project managers look good and become successful, but they need to listen and understand the roles and responsibilities of quality. The quality factors are essential and very important to understand. The Effective Methods for Quality Assurance directions will provide assurance that meet customer requirements before any thoughts about a hurry-up delivery. Stay the course and do not deviate from the required and compliant plans. Before the delivery of products to customers, it is important to complete the following:

- Requirements and data verification and validation are complete.
- Release products are compliant and ready for delivery.
- Necessary reviews and audits and all action items are closed.

12.4 Balance Processes and Understanding Required Tools

The balance processes and understanding required tools are very important for management and team members and may lead to realize that companies are working as effectively as they can and changing a tool is part of the solution. There are good and bad ways to select a tool and how you use it, and there are risks when you focus first on tools before considering the problems being reviewed.

There are many technical choices to balance processes with tool selection that can be used, but a successful experiment depends on having the right evaluation criteria.

A good way to understand problems and requirements is to think about your current process and the process you want to have. You can then evaluate how well a required tool supports the new and balanced processes.

Management and team members often assume that they have a clear understanding of their problems and don't challenge that assumption. A brief analysis of the problems may lead to all solutions that are worked and doing nothing may not be the right approach.

Being inflexible with company processes is another risk and could find a tool that is close to what you need but which does not do things exactly as you would want. In these tasks, you may be tempted to extend the tool in some nonstandard way. If the tool is widely used to solve a fairly standard problem, it can be worthwhile to reconsider the approach, and the problems may not be as unique as you think. Extending or adapting a required tool will fit your nonstandard needs that can lead to extra work and eventually provide proper technical support.

There are many possible inputs and decision processes work to be compliant. The key to successful decisions are to ground in the technical, process, and business needs rather than hype or mess with unsupported assumptions. Examine whether companies can fit process into the tool supports instead of adding complexity. Quality Assurance representatives can help find many customizations to consider whether problems really differ from other issues and concerns. By starting with your process goals first, you can do a better job of evaluating and selecting understood and required tools. This will lead to better decisions and, in the end, lower the costs of businesses.

All required tools have a place in the choice of what type of tools to use is almost as important and understood as the choice of the tool for company use. Management and team members need to take a proactive approach to be based on where they are in application development tasks and activities. When in the development life cycle for companies, is most likely to discuss and utilize the tools required to support the development and requirements.

Further Reading

Cassidy, A. 1998. "*A Practical Guide to Information Systems and Strategic Planning*", Boca Raton, FL, CRC Press. ISBN 1-557444-133-7.

13
Military Aerospace and Defense

Having been a software engineer for 31 years, I was honored to work with Hercules Missile Defense and the Boeing Company for Military Aerospace. Working with many programs for the Boeing Company was a challenge, and as a software engineer, I wanted to make sure that I was doing everything right and always meeting important Quality Assurance requirements. The term "Quality Assurance First" is often used alongside project management, engineering, and primary program functions. Due to the sometimes-catastrophic consequences a single failure can have for human lives, the environment, device we must know that Quality Assurance plays a particularly important role here for military aerospace and missile defense. It has organizational and program developmental independence, meaning that it reports to highest management levels and stands on an equal footing with program and project management and embraces all the customer's point of views.

13.1 Hercules Missile Defense Program

In 1985, my beautiful wife Jana worked with the Hercules Missile Defense company in Clearfield, Utah. I graduated from Weber State University in 1985, and she got me a job working at the same company she worked at. I was a software designer and programmer for the Hercules Missile Defense Company and was able to work all software programming activities with team members who had the experience and knowledge to make sure all software development was compliant and ready for test sites. Working with quality engineers, Software Configuration Management (SCM), and test engineers was good to know that the software was ready for use. Missile guidance experts provided expertise and a good understanding for what is needed from our software team to ensure nuclear missiles are reliable and protective for our country.

13.2 B2-Stealth Bomber Program

In 1987, I had the honor to move to Seattle, Washington, with my family and work with the B2-Stealth Bomber program as an SCM engineer. At times it was a challenge to make sure that all changes to software were implemented correctly and that all software builds were successful for testing and ready for Avionics implementation. The main company for the B2-Stealth Bomber program was Northrup-Grumman. It was great working with this company and coordinating all activities with them. I was working with software designers and programmers, software quality engineers (SQEs), systems engineers, test engineers, configuration management representatives, and numerous suppliers. As a lead engineer, I worked with the Boeing Management, who did everything right and was always great to work with.

13.3 F-22 Raptor Program

The Lockheed-Martin Company was able to provide the F-22 Raptor as the greatest generation advancement in fighter-aircraft capability. The Boeing Company was a contractor to develop Avionics software development and delivery to Lockheed-Martin after integration lab tests and flight test capabilities. The F-22 Raptor mission provided air superiority for air warfare and direct fire to any target within their range. In 1992, I was able to work with the F-22 Raptor program as an SCM Technical Lead engineer. The software engineering team was great to work with and was also able to work with SQE facility to ensure that Quality Assurance was used as an oversight to all software development activities for test and avionics. The F-22 Raptor penetrates the adversity airspace as a stealth jet fighter and has a small radar that makes it invisible to the enemy radar, and the speed is classified to be faster that 1,800 mph. Greater range allows the F-22 Raptor jet to control more airspace and will spend more time in battle areas. I want you to know that the F-22 Raptor program was my best program to work and was able to document SCM and software quality plans (SQPs) and implemented new and current tools (ClearCase and ClearQuest) and was always able to support the F-22 Raptor program at many Air Force Bases (AFBs) to ensure Boeing and supplier software worked and was ready for buy off by Flight Line personnel. After many years working as an SCM Technical Lead, I was able to work as an SQE. It was great to work on Quality Assurance knowing that the software was ready for integration labs and flight tests. I worked with the F-22 Raptor program from 1992 to 2016.

13.4 Advanced Systems Program

In 1995, I was interviewed by Boeing program managers from Denver, Colorado, to help put SCM engineering builds together and working for advanced systems at a military Air Force Base (AFB). I worked in this program for 8 years as an SCM engineer and then worked as an SQE to provide oversight to all software development, testing, and Quality Assurance buy offs. Conducted reviews and audits ensure that software development was working correctly and ready for implementation and use.

13.5 Airborne Early Warning & Control Program

After working in Colorado, I was contacted by the top manager of the Airborne Early Warning & Control program (AEW&C) in Australia. I was relocated back to Washington State to work as an SCM lead and to put a team together for the AEW&C Australia program. There were new software tools (ClearCase and ClearQuest) that were to be used for software development. The team I put together was great in the implementation of these two software tools, and we were successful in all software development activities on the AEW&C program. The management team was great to work with and was very helpful to my team, and working with the Australian engineers was also great. Traveling to Australia was a great adventure, and working with the AEW&C suppliers was important to ensure that all deliveries were compliant. In 2009, I was interviewed by the Boeing Software Quality Manager to work a new program for the AEW&C South Korea program as an SQE. I was expected to document an SQP, perform reviews and audits, First Article Inspections, Functional Configuration Audits (FCA), and Physical Configuration Audits (PCA) for the South Korean program. All SQE tasks and activities were accomplished and worked well for the program and were used in South Korea to ensure that all software activities were working and ready for use in test integration labs and flight tests. It was great working with the Boeing software team while in Washington State and also working in South Korea for 3 years with the South Korean military and a great contractor supplier from Seoul, Korea. When Boeing made the final delivery in 2012, the South Korean military and government appreciated the work that was accomplished and completed.

13.6 P8A Poseidon Navy Program

The final Boeing program I worked for before retirement in 2016 was the P8A Poseidon Navy program in Washington State, working as an SQE, where many reviews and audits working with the P8A Systems Engineering team and management were performed. I was also working on the Boeing Flight Line with hardware and test engineers for buy offs of software to be ready for use on the Boeing Flight Line for all flight tests to be conducted. Also, I worked with numerous Boeing suppliers on P8A and conducted FCA and PCA activities for American and Indian P8A planes for delivery. It was an honor to work with the P8A Navy program. My father (Louis Summers) was in the Navy during World War II, and I always thought about my father during my work with P8A Poseidon.

13.7 Quality Assurance Conferences

As an SQE, I had the opportunity to speak at Boeing conferences. The topic was "Ensure and Implement Quality Assurance in All Boeing Programs." My Boeing coworker Earick Gamble was also a speaker, and he covered the importance of software metrics needed to ensure all engineering tasks and activities were compliant and working per plans, procedures, requirements, and directions. We were also able to speak to a NATO organization to make sure Quality Assurance was implemented and working for all programs. Also, as a technical lead engineer, I was able to talk about the important support I was able to provide and perform for all engineering teams and management and emphasize the importance of Quality Assurance actions in all engineering activities.

13.8 Quality Assurance Technology

Quality Assurance technology and staffing provide military aerospace, and missile defense provides business solutions that are corporate Quality Assurance solutions supporting Automation setup and product and process analysis. Most of our Quality Assurance technology solutions are developed at product and program management, system analysis, application development, data analysis and engineering. Quality Assurance and managers provide effective product support, technical documentation, and plans.

Quality Assurance services and solutions outsource the military aerospace and missile defense product functional testing needs. Quality Assurance technology helps define and design templates that would best fit organizations' quality and control processes.

It was an honor to work with programs for military aerospace and missile defense during my 31 years to ensure that all Quality Assurance and software engineering processes provided the right way of protecting the United States.

Further Reading

Military and Aerospace Electronics. 2018. Vol 29. ISSN 1046-9079.

14
Quality Assurance Plans

Quality Assurance (QA) plans provide details of the overall approach to QA activities and how the project defines, implements, and assures quality for companies, institutions, military programs, and successful business processes. The QA plans define the acceptable level of quality, which is typically defined, and describe how programs and projects ensure the level of quality in its deliverables and work processes. Quality Management plans apply to project deliverables for programs and project work processes. QA representatives along with program and project teams work together in developing a QA plan designed to document the processes and procedures for assuring quality throughout the course of development and implementation. The QA plans comprises a list of attested actions developed to customer satisfaction with products and services, resulting from a program and project that has four distinct executions:

- Planning
- Actions
- Verify
- Execute

14.1 QA Communication

The QA Plans provide communication for project teams, including the managers, developers, test analysts, technical writers, functional analysts, and users to describe the structure of the organization responsible for QA. Management services is responsible for the process component of QA, and the evaluation of the product should be done within projects and by joint customer/developer reviews. The following organizational items should be included or referenced here:

- Program management, program and project executive offices.
- Program and project team organization charts.

- Organizational plans and documents to include any other teams/organizations participating in QA processes.

QA Plans reference standards and guidelines expected to be used on programs and projects to describe how processes determine compliance with these standards and guidelines. Always document relevant artifacts by reference and refer to the following standards and guidelines that may be relevant to the QA Plans. Artifacts are pieces of information that are produced, modified, or used by processes to define an area of responsibility, and are subject to version control.

Documenting all quality records and documentation will be maintained and established to provide evidence of conformance to requirements and to the effective operation of quality management systems. Documentation includes artifacts that are maintained as evidence of product life cycle practices that include baseline work products, summary of work product reports, and more.

QA Communication is an important factor in the improvement of companies, institutions, military programs, and successful business processes. Communication using visual management goals motivates people to commit to change, by showing expected benefits and early results.

Many companies, institutions, military programs, and successful business processes are run by technical people in a technical environment. Often communication is undervalued and underestimated and perceived as difficult. If QA gets started and provides learning along the way, it will get better and require managers, team members, and employees to do something that helps a lot to let them know what you expect from them. There are several ways of communication that will show results using communication tools and techniques that will serve different purposes and are used as a balanced mix of plans and electronic communication needs.

Even more important is the face-to-face communication for the improvement under many changes that help to understand those changes and to build trust to showing up and being visible. Possible forms of communication are as follows:

- Kick-off meetings for management, team members, and employees.

- Training and coaching to ensure understanding of all needs.
- Conduct team meetings daily or weekly.
- One-on-one meetings are important.

Many questions are asked on communication importance and the patterns of communication with companies, institutions, military programs, and successful business that a QA representative way that is talked, email, voice inflection and words selected to determine how everyone relates to management, team members and employees. When asked how to improve communication skills, increase your effectiveness as a QA representative and as a team member.

Communication ultimately determines success and does one of the two things:

- It solves problems or creates them.
- Changes your team, for better or worse.

14.2 QA Plans Provide Education and Learning

The purpose of QA plans provide education and learning by ensuring a level of quality in all products and services to be able to build a positive plan for reliability and consistency. QA must be used in the development of products or services that ensure a level of quality in production and documented in QA plans. Also, QA plans are referred to quality control (QC) to encompass the processes and procedures that systematically monitor different aspects of a service, process, and detect and correct problems that fall outside of established standards or requirements.

Most companies, institutions, military programs, and successful business processes utilize the QA plans to form QA activities in production to development companies and may even be represented by distinct departments or divisions that focus solely on QA issues. A QA representative must have strong skills in a variety of categories: engineering and technology, verbal and written communication, problem-solving, and practical skills like exceptional documentation and the ability to document QA plans.

Analytical skills are important to know how to write and document QA plans to figure out how processes are to be implemented and used. Some QA representatives are naturally better at these types of activities, and it is possible to improve *analytical skills* with practice and activity. QA representatives should have

- An understanding of QA methodologies, tools, and processes.
- Knowledge and working experience in QA.
- Ability to write and document QA plans.

To work in QA technology, there are needs for strength and manual agility to work in QC and technology. QA technology representatives often receive educational training and to increase more advanced training in QA and control management.

There are programs that aim to teach management theories and statistical concepts along with QC topics. The courses might include measurement capabilities, systems engineering, productivity and manufacturing, and process planning. You could choose to advance in quality systems management, in which you would conduct independent reviews, audits, and evaluations to research in QC and reporting results. After working in QA for control and technology, you might consider getting certified to display your proficiency and receive an American Society for Quality award by going to website (*asq.org*). You must meet education, experience, quality, and testing requirements to achieve certification.

14.3 Improving QA Plans

Assign a high priority to improving QA plans for which a failure would have a big impact on documented plans. You don't want a critical plan to fail days before the release date when it's too late to address the issue. Track all efforts spent on to better predict future workloads for companies, institutions, military programs, and successful business processes, and in this way, you can plan the schedule with far greater confidence.

Improving QA plans allows fully archive processes for projects upon completion. Archived processes are protected from modifications and enable you to audit, review, and evaluate past results with confidence. Strong quality results provide protection for ideal choices for teams working on products and projects as well as in regulated companies, institutions, military programs, and successful business processes.

Reporting sections for improving QA Plans make it easy to generate comprehensive project reports, track the coverage of activities, references, and defects as well as many additional metrics and statistics. All built-in plans are highly configurable to adjust them for needs and to cover the scopes required.

QC and QA must be integrated into every step of the development process in QA Plans. Thus, undertaking checks and procedures at every stage of estimation and document development is an open and transparent inventory process, using multiple review processes, and providing communication and feedback across the participants that are all part of QC and improvement. The plan also contains information feedback loops and provides corrective actions that are designed to improve the plans over time.

QA Plan is an important element for quality and QC. The plans should, in general, outline QA/QC activities that will be implemented from its initial development through final reporting in any year. It should contain an outline of the processes and schedule to review all source categories and is an internal document to organize, plan, and implement QA/QC activities. In developing and implementing QA plans, it may be useful to refer to the standards and guidelines published by the International Organization for Standardization (ISO), including the ISO 9000 series.

QC should occur throughout the inventory development and document preparation. QA/QC is not separate from, but is an integral part of, preparing the inventory. The QA plan itself is intended to be revised and reflect new information that becomes available as the program develops, methods are improved, or additional supporting documents become necessary. The QA plan must be realistic and comprehensive in nature to achieve QC standards required by management. If there is a failure to understand customer's expectations, the QA plan will fail short effecting customer service. The key to developing a sustainable QA plan involves management, team members, and employee's accurate communicative dialogs with all programs and project capabilities.

QA representatives must determine roles and responsibilities to ensure effective outlined QC objectives. Management, team members, and employees should own the tasks assigned and with ownership to take responsibility for both success and failures. Clearly designate roles and define those roles with action plans documented. Verify actions by getting on the same page with other business plans. Input from other programs and projects will help assure the objectives set forth that are achievable and presents the best, most cost effective and make sure it aligns goals with all other plans to assure continuity of all tasks and operations.

14.4 Summary

The QA representatives must review and assess the QA plans on an annual basis and provide advice for improvements. The QA plan must be realistic, yet comprehensive in nature to achieve QC standards for requirements to understand program and project expectations and not fail short, effecting plans, processes, and customer service.

Further Reading

Broughton, R. 2008. "Quality Assurance Solutions".
Doherty, G. 2012. "Quality Assurance in Education". doi:10.5772/32434.
Endean, M., Bai, B., & Du, R. 2010. "Quality standards in online distance education". *International Journal of Continuing Education and Lifelong Learning*, *3*(1), 53–73.

15
Quality Assurance for Customers and Suppliers

Quality Assurance for customers and suppliers should always be established for documented plans and procedures for use to maintain effective quality system plans and procedures for companies, institutions, military programs, and successful businesses. Instructions will ensure conformance to customer and supplier requirements that complies with ISO 9001, AS9100, or AS9120 appropriate to the type of product/service being delivered. Any changes to quality plans and procedures that affect the quality of delivered products or services must be approved by management, team members, and employees, and should include a Quality Assurance representative before implementation.

15.1 Customer and Supplier Requirements

Customer and supplier requirements must establish and maintain documented plans and procedures that define the control of all drawings, documents, and digital data that relate to the contract or purchase order requirements. The responsible for the control of all documented plans and procedures are acknowledged for signing and returning the purchase order that will accompany all documentation. All documents initiated by the customer and supplier shall be reviewed and approved by authorized personnel prior to issue with current revisions available at locations where processes that affect product functions are being performed. The customer and supplier shall maintain an effective quality plan and procedure to ensure that the purchased product conforms to specified delivery and purchase requirements.

An approved customer and supplier list must include the scope of approvals and shall be made available upon request. Products procured for use on a contract must be inspected and tested to ensure that all requirements of the applicable material specifications as shown on drawings and/or the purchase order are met. The customer and supplier shall flow down the purchase order and quality plan requirements to direct and ensure that requirements are met.

15.2 Process Control

The customer and supplier shall plan the production and inspection processes that directly affect Quality Assurance and ensure that these processes are carried out under controlled conditions. The controlled conditions shall include the following:

- Documented plans and procedures will define the production, inspection, and delivery, where the absence could adversely affect quality.
- Compliance with all documented plans and procedures show requirements.
- Standards for requirements provide workmanship in the clearest practical manner.
- Accountability for all product and evidence that all inspections have been completed in sequence as planned and authorized.

Processes required by contract shall have an adequate plan and procedure for special processes used in the product life cycle performed at the facilities. It is the customer and supplier's responsibility to utilize approved facilities for special processes per documented requirements. These control measures shall include the following provisions:

- Customer and supplier must ensure special processes approved.
- Special processes shall be performed by accredited processors.
- Customer and supplier shall provide a list of special process suppliers levels at First Article submission.
- Conformance from special processors shall be retained by the Control of Quality Records.

The customers and suppliers shall establish and maintain a process for control of inspection and testing activities to verify that the specified requirements for the product are met. The required process steps required verification and/or validation activities with associated records establishing product realization to be detailed in the Quality Assurance management system. First Article Inspection (FAI) shall provide production and inspection records to verify acceptance of the configuration and performance of the first item submitted for the following categories:

- A new revision to the configuration of the product.
- As specified on Planning Policy Guidance (PPG) Aerospace purchase order.
- If a change in process, tooling, or location has occurred.

FAI activities shall be performed in accordance with Aerospace Standard AS9102, documented on appropriate forms. The FAI report shall accompany the shipment when any of the earlier conditions occur. Incoming products shall not be used or processed until it has been inspected or otherwise verified as conforming to specified requirements. The customers and suppliers must check test reports against specification requirements. If the incoming product is accepted on the basis of certifications or tests reports, periodic scheduled data validation shall be conducted on samples of products with significant operational risk and control. Control of quality records shall maintain adequate records of inspection, test, calibration, and Quality Assurance activities. The records shall provide objective evidence of the operations performed. The records shall be suitable in format, accuracy, and detail to permit analysis by management and a Quality Assurance representative for the initiation of specific corrective actions. Records may be in the form of hard copy or electronic media as documented in Chapter 1.

15.3 Internal Audits

Internal audits shall be conducted for Quality Assurance systems to ensure compliance. Methods of documentation for audit check sheets and recording of audit results are at the discretion of the customer and supplier. The following results:

- Any deficiencies identified shall have a corrective action plan documented.
- Audit Quality Assurance representative shall be independent of supervision of the area(s) audited.

- Management with executive responsibility shall review the internal audit results.
- All internal audit records shall be retained in accordance with control quality records available for reviews.

When it is time to perform audits for customers and suppliers, there will be questions that provide the necessary feedback on how they can improve their processes. The emphases of these disciplines are the key roles for performing effective audits.

Many companies, institutions, military programs, and successful businesses have found issues and concerns to effectively perform internal audits. The internal audits are beneficial when it is time to prepare for informal and formal audits during critical program or project schedules. It is important to provide detailed information and the steps required to perform effective audits. Artifacts (i.e., plans, processes, procedures, test reports, data, etc.) are required for these specific internal audits. These audits ensure that informal and formal audits are ready to be conducted and performed properly and in compliance to required requirement standards. The important factor for audits to be successful is you must have a combined effort of a process model and a Quality Assurance management system working together.

The requirements phase is the process of defining the data for a system to satisfy specified system requirements. System documentation conveys requirements, design details, capabilities, limitations, and other characteristics to show reliability. Audit is the process to verify that specified requirements are in compliance and control.

The test phase is a period of time in the product life cycle during which components are audited and evaluated, integrated to whether or not requirements have been satisfied and compliant. Test data and associated test plans and procedures exercise program paths to verify compliance to specified requirements. The test plans describe the approach intended to identify the items to be tested, the testing performed, test schedules, reporting requirements, evaluation criteria, and risk management. Test procedures provide detailed instructions for setup, operation, and evaluation results.

15.4 Summary

The product life cycle phase is the framework that provides an understanding of all processes. The relationship between the process model and Quality Assurance ensures that delivery of products is compliant and meets requirements. Performing effective internal audits during the phases of the product life cycle will benefit future preparation towards informal and formal audits.

Further Reading

Wellins, R. S., Scaff, D., & Shomo, K. H. 1994. "*Succeeding with Teams*", Minneapolis, MN: Lakewood Books.

16
Supporting Software Configuration Management

Quality Assurance (QA) representatives should always support Software Configuration Management (SCM) processes that are documented in plans and procedures for managing changes. The benefits of change management processes follow change decisions to ensure companies, institutions, military programs, and successful business processes maintain consistency for configuration baselines. SCM is a process for establishing and maintaining consistency of a product's performance, functional, and physical attributes with its requirements, design, and operational information throughout its product life cycle. The SCM process is widely to manage changes throughout all the activities being conducted by employees that have the knowledge to perform. SCM is the practice of handling changes to maintain its role for complete times. SCM implements the plans, policies, procedures, techniques, and tools that manage, evaluate proposed changes, track the status of changes, and support documents as the required changes.

SCM plans provide technical and administrative direction to the development and implementation of the procedures, functions, services, tools, processes, and resources required to support all activities. During development, SCM allows management to track requirements throughout the product life cycle through acceptance and operations and maintenance. As changes inevitably occur in the requirements and design, they must be approved and documented, creating an accurate record of status.

The SCM process for both hardware and software configuration items (CIs) comprises disciplines as established. These disciplines are carried out as policies and procedures for establishing *baselines* and

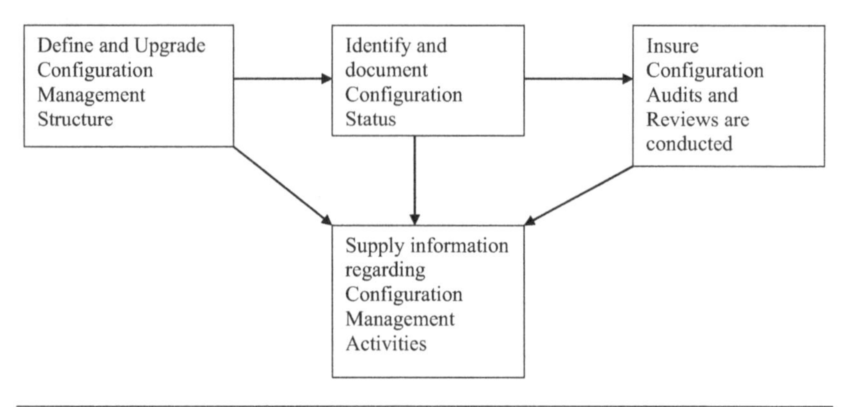

Figure 16.1 Software Configuration Management.

for performing a standard *change-management* process of activities for release management and delivery. Overview of Configuration Management is provided in Figure 16.1.

16.1 QA and Configuration Management

QA representative team must manage the quality of entire programs and projects for SCM. They ensure that the internal processes as well as documentation for plans and other media items fulfil customer needs. The team is made up of experienced personnel who have hands-on experience in ensuring that products and services are delivered on time and complies with established standards.

Our QA management services include

- QA and training.
- Quality control expectations.
- Quality management focus.
- Quality audits, reviews, and evaluations.

SCM is crucial in ensuring practices and the control of releases to companies, institutions, military programs, and successful businesses. Some SCM systems may contain thousands of program files and codes, which change as the development work progresses. The integration of these programs with other components further complicates the versions of the entire SCM system. Development teams identifies and manages all CIs and control statuses in cases where these items are needed to be changed or modified, such as code changes and document changes.

Under the oversight QA, SCM may have access to all data, process documentation, and products and outputs associated with management, team members, and employees. QA representatives must review all sources for completeness and accuracy before distributing and releasing to customers that are under configuration control and receives, distributes, and applicable customer outputs under configuration control for companies, institutions, military programs, and successful businesses. The SCM process provides a system to identify, store, retrieve, and distribute data with protection to ensure qualification of all personnel authorized to perform SCM functions and tracks the changes that are made to data items back to the original source.

In the field of QA, inspection of SCM development processes lead to an improved use of automation, and functional testing and implementations of changes are considered to be effective executed activities. The results of change activities can be separately provided to QA for quality inspection as such for functional testing. This is possible due to the administrative support for product and configuration administration with SCM tools that are used in the SCM database.

During programing for particular change insures configuration of the development for integration with change are provided. To improve the integration of a change, SCM provides the development team with an integration configuration to emerge the new code in order to reduce the overall effort for the execution of regression tests for changes that intends to subdivide the regression testing. In addition to the support for QA inspection support for SCM, QA also is in the field of quality planning and quality control. The change management database will also provide planned activities with Quality Assurance to related activities to be planned.

Test cases can be selected from the test database for regression tests or reviewer for code reviews that can be included in the plan at determined times. Based on the results of the QA inspection activities, the evaluation of problem reports is part of the change management database to ensure that quality control can be done. Development parts of the SCM database with increasing bug rates can be detected, or activities, which are susceptible to mistakes, can be identified. This allows the selective and planned use of appropriate QA inspection activities.

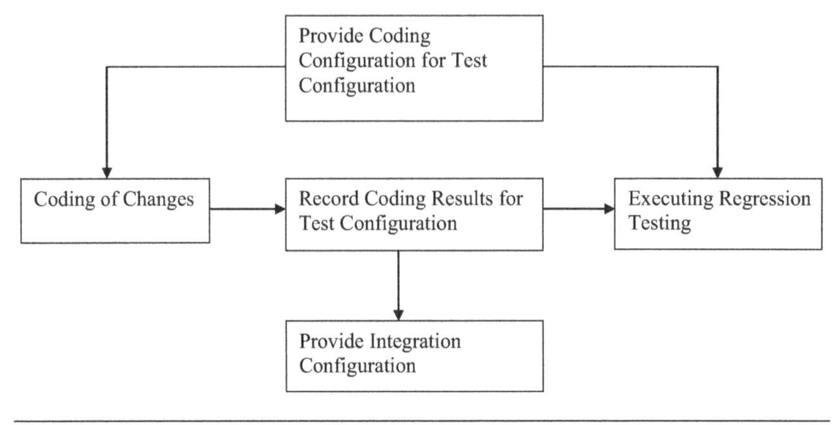

Figure 16.2 SCM-tool control.

Interaction of QA and SCM for Quality Inspection is proved in Figure 16.2.

The change management process is monitored by QA representatives multiple times, and assessments to ensure processes are assessed and approved. The change request (CR) is used to verify that the change control processes are in place and operate effectively. Issues identified during reviews of the configuration management process will be monitored by QA to ensure that methods are approved for use and include creating plans, procedures, and milestones requested by management. Measurements are used to determine the status of control implementation and effectiveness for SCM activities and processes as stated in the following two items:

- Control weakness and problems.
- Control significant deficiencies that have issues and concerns.

16.2 SCM Planning

SCM planning required documenting and planning for management, team members, and employees for responsibilities and requirements as stated later:

- Training requirements for SCM employees.
- Meeting guidelines and including a definition of procedures and tools.

- Baselining processes used for delivery.
- Configuration control and configuration status for all development activities.
- Audits and reviews along with evaluations that could be conducted by QA.
- Customer and supplier SCM requirements.

SCM identification consists of setting and maintaining baselines, which define the architecture, components, and any developments at any point in time. It is the basis by which changes to any part of SCM are identified, documented, and later tracked through design, development, testing, and final delivery to customers. SCM incrementally establishes and maintains the definitive current basis for Configuration Status Accounting (CSA) and its CIs throughout their product life cycle (development, production, deployment, and operational support). Configuration Control includes the evaluation of all change requests and change proposals and their approval or disapproval. It covers the processes of controlling modifications to the design of hardware, firmware, software, and documentation.

CSA includes the processes for recording and reporting CI descriptions for hardware, software, and firmware and reviews of departures from the product life cycle baselines during design and production. In the event of suspected problems, the verification of baseline configuration and approved modifications can be quickly determined.

Configuration Audits and Reviews is an independent review of hardware and software for the purpose of assessing compliance with established performance requirements for product baselines. These audits and reviews verify that all configuration documentation complies with the performance characteristics before acceptance into product-level baselines.

The information about effective processes is necessary for management, team members, and employees and should be made available by SCM. Changes are documented and the updated status for the information should always be checked. Updated information relating to the CIs is continuously available within the Configuration Management Database.

16.2.1 Software Configuration Management

The responsibilities of SCM are as follows:

- Implement the SCM program and SCM requirement baselines.
- Establish and maintain a CR tracking database based on the Configuration Control Board.
- Develop and document the SCM plans and operating procedures.
- The Engineering Review Board (ERB) will prepare and distribute its agendas and minutes, recording status of CRs affected by ERB reviews and approval.
- Produce and distribute CR status reports.
- Coordinate CRs with individuals responsible for implementation of approved changes and resolve and fix all CR problems.
- Make sure QA representatives are included in ERB meetings.

The role of the SCM may be filled by an individual responsible for ensuring all CRs steps were documented as completed for ensuring that change request steps were documented as completed.

16.3 Change Request Management

Change Request Management (CRM) is for recording, tracking, and reporting of request from users to change and implement updates for companies, institutions, military programs, and successful business for maintaining and applying configuration baselines. It includes decision making to decide what CRs are needed for updates needed. It is important for CRM to provide complete change solutions and recording to maintain tracking CRs to determine the level of product development and QA to provide status to management.

16.4 Change Requests

CRs are known terms for requests from designers to change artifacts and processes. Documented CR information that relate to any issues and problems that are needed to be a solution that needs to be fixed. The CR processes are related to internal organizations and are divided into two major categories:

- High-level requests for fixing any problems.
- Fix defects that are discovered during development activities.

High-level requests specify new changes needed to change system development, and defects are a flaw and can be any kind of issue you want tracked and fixed, and these two categories represent CRs that are needed to address issues and problems that need to be fixed by designers and development representatives. Once CRs are defined, it is important to track and define the information needed to record what changes are to be used. If you are tracking defects, it is important to record what is submitted as a defect when submitted for resolution during the life cycle phase. CRs include the following six stages:

- CRs that are submitted to change any systems and development.
- All CRs need to be evaluated and prioritized.
- Evaluations provide decisions to be scheduled for implementation.
- CRs that are implemented are produced to reflect change.
- Verification of CRs meet the requirements for fixing defects.
- On the completion of the CR, the requested is notified.

During submission of CRs, all requests are recorded and enhanced for all collection to be collected. Defects come from a wide variety of changes and should be recorded and resolved to ensure that issues and problems are fixed. Questions could be asked on how defects are noticed and the severity of who discovered the defect. The key data recorded for defects provide the identity of customers having problems and that is where evaluations should be conducted and possibly involve QA to ensure that quality is implemented. Evaluations for submitted CRs are conducted by the engineers working for the updates required. My feeling is that a QA representative should work with the engineers to ensure that updates and fixes are completed and working and know that this action is needed for process evaluations. During evaluations of CRs, you are looking at the importance of what customers need to continue, knowing that the changes made will be compliant and working.

The decision stage is when engineering is ready to implement the CR and enhance everything that is handled correctly.

16.5 What is ClearQuest?

ClearQuest is a software change management tool that helps improve software development productivity while accommodating the changes, processes, and tools that best fit programs and projects on software teams. This software provides tools and processes that allow control of changes provided by software developers. Rational ClearQuest is a comprehensive change tracking system for software development environments and can manage all types of CRs, including defects, product changes, issues, requests for new activities, and documentation for plans and procedures for updates and changes. Rational ClearQuest uses a flexible workflow process that can be tailored to specific needs and to phases of the software development processes. Companies can use Rational ClearQuest to automate and enforce development and to manage issues throughout the project life cycle. Before software engineers create CRs, development usually sets up an action for applicable programs and projects. The description of the CR process for records is stored in a database that includes CRs and user interfaces.

Rational ClearQuest comes defined schemas and effective configurations, and once the schema is deployed, engineers can begin submitting and working on CRs.

This tool can integrate with other products, including requirements, development, build, test, deployment, and management tools, which help ensure a faster response to change.

Integration with test management tools helps unify development and testing activities, from planning to results, for improved QA. The CR is from the software development process to change something in a product or process that includes defects and a request for product enhancement. To initiate CR, engineers complete a form and submit it as a record that moves through a series of information and actions. The CR is its current status and includes being submitted, assigned, and closed.

Figure 16.3 shows how a defect moves through the CR process:

A development manager may assign to a developer in an open state until the developer makes a decision about the defect. If the developer chooses to fix the defect, it must be given a resolved state. At that point, a QA engineer looks at it and validates that it is actually fixed and

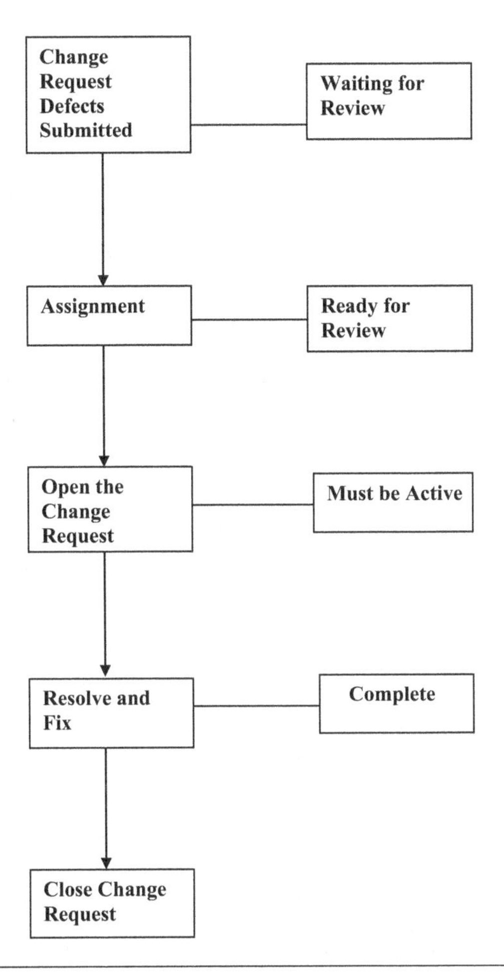

Figure 16.3 CR defects.

working. Then, the Quality Engineer changes the state from resolved to closed. Software engineers can manage the CR environment from the Web for ClearQuest or from Windows. Administrators configure and manage that environment using various administration tools.

After working for years and being an SCM engineer, I was able to work ClearQuest which is a Rational tool. Working with many programs for Boeing, I put an SCM team together, and we implemented and installed ClearQuest to handle all CRs provided by software development teams. ClearQuest works with ClearCase, which is another SCM tool for software development teams to provide updates and other needed software capabilities.

16.6 Software Builds and Release

For software builds and release, it is an important job for the SCM engineers to take the work of many teams that can be tested per requirements. There is integration that is merged and assembled for the software teams. I would like to talk about Rational ClearCase now, and how the tool covers builds, baselines, and releases.

Rational ClearCase is a *computer software* tool that supports SCM to the handling of *source code* and other *software development* activities. It also supports design and data management of artifacts, enabling hardware and software development. ClearCase includes *revision control* and works on the basis of SCM in businesses accommodating projects with software designers and developers.

ClearCase can accommodate large numbers of files, and supports branching, labeling, and versioning. In ClearCase terminology, the individual database is called a Version Object Base (VOB). A set of interfaces with accompanying tools are used to manage the physical database system, which requires specific database administrator skills. Under ClearCase software is controlled by associated configuration specification and stored internally in a text file and compiled before use that specifies what element versions (files or directories) are to be displayed in a view. To determine which version, if any, of an element should be visible, ClearCase traverses the configuration specification line-by-line from top to bottom, stopping when a match is found and ignoring any subsequent rules. I have always involved QA representatives to help with the processes to ensure plans and procedures are being followed and attending ERB meetings to understand ClearCase and ClearQuest information.

Integration brings together developed changes to form a test of software systems, and it can occur at many levels of completion. There are integration tasks required to manage all software development efforts relevant to SCM to merge and put together integration. Merging integration involves changes made by software teams that have been modified and necessary to combine or, in ClearCase terms, merge all changes. It should be clear that merging integration requires the understanding of changes being made and to be introduced to the software for performing merge integration and to resolve any conflicts. There is integration information based on software engineering development to perform assembly integration. The software teams work development for software to make sure that the software is ready to be integrated and tested. Major software has two levels of integration: the first is done by project teams involving QA and the second level brings all testing to be accomplished and used.

Software developers perform internal builds, along with all other work related to ClearCase work in separate and private views. As described in programming software, each role provides a complete environment for building the software that includes a particular configuration of software versions and in a private work area in which you can modify source files.

As a build, environment building software should never disturb the work in another build of the same program at the same time. However, when working in a dynamic view, you can examine and benefit from the work done previously in another dynamic view. A new build shares files created by previous builds and sharing saves the time and disk space involved in building of new objects that duplicate existing ones.

ClearCase includes tools for listing and comparing past software builds when VOBs constitute a globally accessible repository for files created by software builds, in the same way that they provide the source files that are used into software builds. A file produced by a software build is associated with each derived object as a Configuration Record that is in use during subsequent software builds to determine what can be reused or shared.

Further Reading

Keyes, J. 2004. "Software Configuration Management" ISBN 0-8493-1976-5 *PACO Technologies, Inc. Archived from the original on 26 August 2016*. "Configuration Management Case Study". Retrieved 28 March 2012. Quality Management Systems 2003 "Guidelines for Configuration Management" ISO 10007. https://www.goodreads.com/book/show/367726.Software_Configuration_Management?ac=1&from_search=true

White, B. A. 2000. "*Software Configuration Management Strategies and Rational ClearCase*", Boston, MA, Addison-Wesley. ISBN 0-201-60478-7.

Appendix A

Acronyms and Glossary

Agile: Agile development describes an approach to software development under which requirements and solutions evolve through the collaborative effort of self-organizing and cross-functional teams and their customer/end user.

Audit: An independent examination of a work product for software or set of work products to assess compliance with specifications, standards, contractual agreements, or other criteria.

Baseline: A specification or product that has been formally reviewed agreed upon and can only be changed through formal control processes.

Benchmarking: Benchmarking is a process of measuring the performance of a company's products, services, or processes and to identify internal opportunities for improvement.

Build: Operational version of a software product incorporating a specified subset of capabilities that informal and formal work products include in multiple configurations.

Capability Maturity Model Integration: Collection of process models and methods for use in new disciplines to be integrated for organizational structures.

Change Control: The processes by which a change is proposed, evaluated, approved, or rejected. Scheduled and tracked.

Change Requests: Request to Software Configuration Management to provide software builds for software systems and use for computer labs and support formal testing.

ClearCase: ClearCase is a computer software tool that supports software configuration management (SCM) of source code and other software development assets. It also supports design-data management of electronic design artifacts, thus enabling hardware and software codevelopment.

ClearQuest: Rational ClearQuest is an enterprise-level workflow automation software tool from the Rational Software division. ClearQuest is configured as a bug tracking system to track complex processes.

Configuration Management: The process of identifying and defining the configuration items in a system, controlling the changes and release of these items throughout the system life cycle, and recording and reporting status of change requests to verify completeness.

Data: A representation of facts, concepts, or instructions suitable for communication, interpretation, or processing.

Defect: The aspect of software development/coding issues to resolve on a timely basis and drive daily software execution.

Delivery: The point in the product development life cycle at which a product is released to its user for operational use.

Design: The purpose of defining the software architecture, components, modules, interfaces, and data for a software system to satisfy specified requirements.

Engineering Review Board: Established for the software Integrated Product Teams (IPTs) to review and disposition changes that affect controlled software and related documentation.

Effective Process: Improve business to increase expectations, technologies, and competition of effective ways to establish continual and business process improvements.

First Article Inspection: The inspection performed to assure engineering requirements and processes have been applied to development and release activities.

Hardware: Physical equipment used in data processing, as composed to computer programs, plans, procedures, and associated documentation.

Healthcare: Healthcare is the maintenance or improvement of health via the diagnosis, treatment, and prevention of disease, illness, injury, and other physical and mental impairments in human beings.

High-Performance Work Team: Teams that drive success through developing and leading high-performance teams to ensure complex tasks are current for the competitive work environment.

Implementation Phase: The period of time in the product life cycle during which work products are created from design documentation.

Inspection: A formal evaluation in which requirements are examined in detail to detect faults, violations of development standards, and other problems.

Management: The administration of an organization and business that includes the activities of setting the strategy and coordinating the efforts of employees, and management may also refer to those employees that help manage an organization.

Metrics: Standard for measuring and evaluation that uses statistics for fixing.

Military and Defense: Military and Defense is a professional organization formally authorized by sovereign states to use weapons to support the interests of the states and country and consists of branches such as an *Army*, *Navy*, *Air Force*, *Marines*, and *Coast Guard*.

Mission: Driving the growth of management, team members, and employees, all businesses through personal and professional development focused on disciplined execution and quality.

Peer Review: An important part of verification and proven activities for effective defect removal.

Physical Configuration Audit (PCA): Identifies the product baseline for production and acceptance of the work product audited. PCA verifies that the "as-built" configuration correlates with the "as-designed" product configuration, and the acceptance test requirements are comprehensive and meet the necessary requirements for acceptance of the production unit.

Policies: A set of policies are principles, rules, and guidelines formulated or adopted by organizations to reach long-term goals and plans.

Procedure: The documented description of a course of action taken to perform activities or resolve problems. Manual steps or processes that need to be followed.

Process: To perform defined instructions during the software and product development life cycle.

Program: A schedule or plan that specifies actions to be taken.

Project Plan: A management approach that describes the work to be done, resources required, methods to be used, reviews, audits, the configuration management, and Quality Assurance procedures to be implemented.

Quality: The totality of features and characteristics of a product or service that has the ability to satisfy the required needs.

Quality Assurance: A planned and systematic approach to provide adequate confidence that all products conform to established requirements.

Quality Control: The tasks and activities for quality control determine which behaviors are good and add value towards goals, knowing that it is important because it helps define risks that organizations can take to help manage organizational support.

Quality Management System: Software industries and software programs who establish, document, implement, and maintain an effective quality management and to continually improve its effectiveness.

Quality Assurance Metrics: Measurement of the degrees to which software possesses given attributes that affect quality and standard for measuring statistics.

Quality Assurance (QA) Plans: QA plans provide education and learning by ensuring a level of quality in all products and services to be able to build a positive plan for reliability and consistency.

Requirement: A condition or capability needed by a user to solve a problem or achieve an objective. The condition of capability must be met by a system to satisfy a contract, standard, or specification.

Requirement Analysis: The process of studying user needs and arrive at a definition of requirements and verification is also performed.

Requirements Phase: The period of time in the product life cycle during which the requirements of a work product, such as functional and performance capabilities are defined.

Review: Informal or formal review of system requirements, software design, software configuration management, quality, test, and required data show compliance to documented plans, processes, and procedures.

Results: To proceed or arise as a consequence, effect, or conclusion.

Risk Management: Process to identify risks and identify an approach to prevent future risks.

Software: Computer programs, procedures, rules, and any documentation pertaining to the operation of data processing systems. It is in contrast with Hardware.

Software Configuration Management: Establish and maintain the work product identification process and control changes to identified software work products and their related documentation.

Record and report information needed to manage software work products effectively, including the status of proposed changes and the implementation status of approved changes. Maintain auditable records of all applicable software work products that help verify conformance to specifications, interface control documents, contract requirements, and as-built software configurations.

Software Configuration Management Plan: Configuration Management Plan for controlling and management of software work products during the phase of a software development program.

Software Development Process: The process by which a user needs translated into software requirements, and transformed into design/code being tested, documented, and certified for operational use.

Supplier Data Requirements List: Track Specification Control Documents, Supplier's Design, Approvals, and Acceptance.

Software Metrics: A standard of measurement of metrics for software performance, planning work items, and measuring productivity.

Software Quality: Features and characteristics of a software product that satisfy need and conform to specifications.

Software Quality Assurance: A planned and systematic approach to provide adequate confidence that the product conforms to established requirements.

Software Tools: Computer tools used to develop, test, analyze, and maintain a computer program and its documentation.

Source Code: Computer programs written in a computer language that requires a translation provided by a computer system.

Systems Engineering: Analysis, requirements understanding, and the importance of software design capabilities. Interfaces are defined externally and internally to ensure hardware and software is compatible for supporting team activities.

Subsystem: A group of assemblies or components or both combines to perform a single function.

Suppliers: Supplier requirements establish and maintain documented plans and procedures that define the control of all drawings and documents that relate to the contract or purchase order requirements.

Testing: The process of exercising or evaluating a system by manual or automated means to verify that requirements satisfy expected results.

Test Report: A document describing the conduct and results of testing carried out for a system or system component.

Validation: Validation demonstrates that the product, as provided to fulfill its intended use.

Verification: Verification addresses whether the work product properly reflects the specified requirements.

Work Product: A product that consists of requirements, diagrams, documentation, and development folders.

Index

A

Advanced systems program, 92
AEW&C, *see* Airborne Early Warning & Control program (AEW&C)
AFBs, *see* Air Force Bases (AFBs)
Agile management model, 8–9, 55
Airborne Early Warning & Control program (AEW&C), 92–93
Air Force Bases (AFBs), 91, 92
AS9100, 2
Audit, 3
  internal, 103–104
  performance of, 3
  recording and reporting, 4

B

Baselines, 107–108
Benchmarking process, 61
Boeing Company, 10, 91
Boeing Flight Line, 93
B2-Stealth Bomber program, 90
Building and maintaining teams, 35
Business impacts, 36–37

C

Capability Maturity Model Integration (CMMI), 65, 72–73
Change control, 110
Change-management process, 108
Change request (CR), 110, 112–113, 115
Change Request Management (CRM), 112
CIs, *see* Configuration items (CIs)
ClearCase, 115–117
ClearQuest, 114–115
Clinical peer reviews, 26
CMMI, *see* Capability Maturity Model Integration (CMMI)
Communication
  needs, 72
  for project teams, 95–97
Compliance verification, 9, 18

Configuration
audits and reviews, 111
control, 111
Configuration items (CIs), 107, 111
Configuration Record, 117
Configuration Status Accounting (CSA), 111
Consistency, for Quality Assurance, 19–20
Control charts, 27
Control of quality records, 102
Corrective actions, with metrics, 56–57
CR, *see* Change request (CR)
CRM, *see* Change Request Management (CRM)
CSA, *see* Configuration Status Accounting (CSA)
Customers
internal audits, 103–104
process control, 102–103
requirements, 101–102
satisfaction, 15–16
Cycle time, 55

D

Data control, 20–21
Decision Analysis and Compliance, 73–75
Defect, change requests, 113, 115
Delivery, of complex products, 7, 87
Design templates, 94
Direction
balance processes and understanding tools, 87–88
effective methods for, 85–87
high-level, 83–84
roles and responsibilities, 84–86
Document action plan, 67, 68
Documentation, 20–21
Driving innovation, to reduce costs, 7

E

Effective process
environment, 66–67
evaluations, 18–19
Engineering Review Board (ERB), 112

F

Face-to-face communication, 96
FCA, *see* Functional Configuration Audits (FCA)
First Article Inspection (FAI), 92, 103
F-22 Raptor Program, 91
Functional Configuration Audits (FCA), 92, 93
Functional Quality Assurance tasks, 18

G

Goals and Process Capability, 77

H

Hardware, 12–13, 111
Healthcare, 26
software and Quality Assurance improves, 28
Hercules Missile Defense Program, 89–90
High-level Quality Assurance direction, 83–84
High-Performance Work Team (HPWT), 11–13

I

Improvement, Quality Assurance
clear-cut objectives to, 37–38
definition of, 38

guidelines, 33
with metrics (*see* Metrics)
preparation of, 33
progress, 31
Inspection, 20, 102, 109
Interfaces, 6
Internal audits, 103–104
ISO 9001, 2, 20

L

Lead time, 55
Lean principals, 8–9, 55
Lockheed-Martin Company, 91

M

Management
company and business, 75
vs. employees, 42
quality, 4–5
services, 95
systems, 73
and team members, 6
upper, 13
Medical peer reviews, 26
Metrics, 49–50
corrective actions with, 56–57
database, 52
measure program and project performance, 50–52
software, 53–54
used for success, 54–56
Military aerospace and defense
advanced systems program, 92
Airborne Early Warning & Control program, 92–93
B2-Stealth Bomber program, 90
F-22 Raptor Program, 91
Hercules Missile Defense Program, 89–90
P8A Poseidon Navy program, 93
Quality Assurance conferences and technology, 94
Mission, Quality Assurance, 2

O

Operations management, definition of, 24
Organization-level policies, 6

P

P8A Poseidon Navy program, 93
PCA, *see* Physical Configuration Audits (PCA)
Peer review method, 12, 13
Performance, Quality Assurance
guidelines, 33
improvement, 38, 49
preparation and accomplishment of, 31–32
progress, 31
Physical Configuration Audits (PCA), 92, 93
Plan implementation, 11–12
Policy, in business process, 5–6
PPQ, *see* Process Performance Qualification (PPQ)
Preventive actions, with metrics, 57
Prime importance, 80
Proactive approach, to Quality Assurance, 9–10
Problem-solving methods, 35–36, 56
Procedure, 21, 25
Process control, 102–103
Process evaluations, 18–19
Process execution, 67–69
Process improvement, 13, 14
Adversity and Concerns, 24–25
commitment to, 27–28
control charts, 27
direction, 65
document action plan, 67, 68

Process improvement (*cont.*)
effective process environment, 66–67
guidance and strategies for, 66
healthcare, 26
measurement performance of, 23–24
procedural steps, 28–29
quality control, 25–26
services, 73, 74
software and, 28
strategic planning for, 27
Process infrastructure, 66
Process Performance Qualification (PPQ), 80–82
Product metrics, 54
Programs and projects success, 71–72
Project management, 45
Project plan, 87

Q

QAF, *see* Quality Assurance Framework (QAF)
QARs, *see* Quality Assurance representatives (QARs)
Quality assessment, definition of, 38
Quality assurance (QA), 1–2
conferences, 94
definition of, 38
excellence, 34–35
interviews, 42–43
methods, 17–18
stability, 22
standards, 21–22
plans, 95
communication, for project teams, 95–97
provide education and learning, 97–98
team, 15
technology, 94
Quality Assurance First, 10–11, 89
Quality Assurance Framework (QAF), 59
benchmarking process, 61
benefits of, 62–63
for companies and institutions, 61
evidence, 59–60
requirements and risk mitigations, 60
Quality Assurance representatives (QARs), 18, 68, 69, 85
conduct questions and answers, 39–40
management mistakes, 41
show strength and never weakness, 41
support employees, 41–42
techniques required, 39
Quality-based metrics, 56
Quality control, 25–26
Quality engineering, 6–7
Quality Engineering Knowledge, 80
Quality management plan (QMP), 43, 95
Quality management system (QMS), 4, 14–15
Quality Performance and Improvement Report, 31
Quality planning, 3
Quality principles, 43
Quality process, definition of, 38
Quantitative process commitments
goals and process capability, 77
perform commitments, 77–78
verification and validation, 78–80

R

Rational ClearCase, 116
Rational ClearQuest, 114
Requirement analysis, 6, 79
Requirements phase, 104

Risk management, 45
activities, 47
effective plan for, 46
projects, types of, 45–46
for Quality Assurance process, 48
Risk mitigation, 46
Quality Assurance Framework, 60
Risk monitoring, 46–47

S

SAE AS9110, 2
SCM, *see* Software Configuration Management (SCM)
Six Sigma project, 62–63
Software builds and release, 116–117
Software Configuration Management (SCM), 107–108
change requests, 112–113
ClearQuest, 114–115
planning, 110–111
Quality Assurance and, 108–110
responsibilities of, 112
software builds and release, 116–117
tool control, 110
Software development process, 12, 53, 114
Software engineering, 10
Software metrics, 53–54
Software quality, 52, 54
Software quality engineers (SQEs), 92, 94
Software quality plans (SQPs), 92
Software tools, 92
Source code, 54, 116
SQEs, *see* Software quality engineers (SQEs)
SQPs, *see* Software quality plans (SQPs)
Strategic planning, for Quality Assurance, 27
Suppliers
data requirements list, 101–102
internal audits, 103–104
process control, 102–103
requirements, 101–102
Systems engineering, 10

T

Team members, 85, 86
Team velocity, 55
Testing, 12, 13, 54, 102
Test report, 103
Tradeoffs, 41

V

Validation, 78–80
Verification, 78–80
Version Object Base (VOB), 116–117
Vision, Quality Assurance, 2
VOB, *see* Version Object Base (VOB)

W

Working framework, 36
Workplace problem, 42
Work product, 18, 78, 79